Annals of Mathematics Studies

Number 126

An Extension of Casson's Invariant

by

Kevin Walker

PRINCETON UNIVERSITY PRESS

———

PRINCETON, NEW JERSEY

1992

The Annals of Mathematics Studies are edited by
Luis A. Caffarelli, John N. Mather, and Elias M. Stein

Princeton University Press books are printed on acid-free
paper, and meet the guidelines for permanence and durabil-
ity of the Committee on Production Guidelines for Book
Longevity of the Council on Library Resources

Printed in the United States of America
by Princeton University Press, 41 William Street
Princeton, New Jersey

Library of Congress Cataloging-in-Publication Data

Walker, Kevin, 1963-
    An extension of Casson's invariant / Kevin Walker.
      p. cm.—(Annals of mathematics studies; no. 126)
    Includes bibliographical references.
    ISBN 0-691-08766-0 (CL)—ISBN 0-691-02532-0 (PB)
    1. Three-manifolds (Topology) 2. Invariants. I. Title.
  II. Title: Casson's invariant. III. Series.
  QA613.W34   1992
514.3—dc20   91-42226

Contents

An Extension of Casson's Invariant

§ 0

Introduction

In lectures at MSRI in 1985, Andrew Casson described an integer valued invariant λ of oriented homology 3-spheres (see also [AM]). $\lambda(M)$ can be thought of counting the number of conjugacy classes of representations $\pi_1(M) \to SU(2)$, in the same sense that the Lefschetz number of a map counts the number of fixed points. (Warning: According to Casson's original definition and [AM], $\lambda(M)$ is half this number.)

More precisely, let (W_1, W_2, F) be a Heegaard splitting of M (see (1.A)). For X any space, let $R(X)$ denote the space of conjugacy classes of representations $\pi_1(X) \to SU(2)$. We can make the identification

$$R(M) = R(W_1) \cap R(W_2) \subset R(F).$$

$R(W_1)$ and $R(W_2)$ have complementary dimensions in $R(F)$, and, roughly speaking, $\lambda(M)$ is defined to be the intersection number

$$\lambda(M) \stackrel{\text{def}}{=} \langle R(W_1), R(W_2) \rangle.$$

$R(W_j)$ and $R(F)$ are not manifolds, but singular real algebraic sets, so it might seem that there is some difficulty in defining the above intersection number. But the condition that $H_1(M; \mathbf{Z}) = 0$ guarantees that after excising the trivial representation $R(W_1) \cap R(W_2)$ is a compact subset of the non-singular part of $R(F)$. Thus, after an isotopy supported away from the singularities of $R(F)$, $R(W_1) \cap R(W_2)$ consists of a well-defined (signed) number of non-singular points, and $\langle R(W_1), R(W_2) \rangle$ is defined to be this number. It is not hard to show that this does not depend on the choice of Heegaard splitting.

It is clear from the definition that $\lambda(M) = 0$ if $\pi_1(M) = 1$. In particular,

(0.1) $\lambda(S^3) = 0$.

3

Casson also showed that λ has the following properties.

(0.2) Let K be a knot in an integral homology sphere and let $K_{1/n}$ denote $1/n$ Dehn surgery on K. Let $\Delta_K(t)$ be the (suitably normalized) Alexander polynomial of K. Then

$$\lambda(K_{1/n}) = \lambda(K_{1/0}) + n\frac{d^2}{dt^2}\Delta_K(1).$$

(0.3) $\lambda(-M) = -\lambda(M)$, where $-M$ denotes M with the opposite orientation.

(0.4) $\lambda(M_1 \# M_2) = \lambda(M_1) + \lambda(M_2)$, where $\#$ denotes connected sum.

(0.5) $4\lambda(M) \equiv \mu(M)$ (mod 16), where $\mu(M)$ denotes the signature of a spin 4-manifold bounded by M.

It is not hard to show that (0.1) and (0.2) uniquely determine λ. Also, (0.3), (0.4) and (0.5) follow easily from (0.1) and (0.2). The proof of (0.2) involves clever exploitation of the isotopy invariance of $\langle R(W_1), R(W_2)\rangle$.

For a more detailed overview of λ for integral homology spheres, see the introduction of [AM].

This paper contains generalizations of the above results to the case where M is a rational homology sphere (i.e. $H_*(M; \mathbf{Q}) \cong H_*(S^3; \mathbf{Q})$, or $|H_1(M; \mathbf{Z})| < \infty$). In this case, $R(W_1) \cap R(W_2)$ contains nontrivial singular points, so finding an isotopy invariant definition of $\langle R(W_1), R(W_2)\rangle$ is more difficult. It can, nevertheless, be done, and one can use the isotopy invariance to prove a generalization (0.2) (see (4.2)), and hence generalizations of (0.3), (0.4) and (0.5) (see (6.5)). For a more detailed introduction, the reader is encouraged to read (1.A) and the beginning of (2.A).

Just as in the integral homology sphere case, the generalized Dehn surgery formula and the fact that $\lambda(S^3) = 0$ uniquely determine λ for all rational homology spheres. In this case, however, it is not very hard to prove that the Dehn surgery formula does not overdetermine λ. In other words, one can use the Dehn surgery formula to define λ and avoid $SU(2)$ representations altogether.

Section 1 contains results on the topology (and symplectic geometry) of certain representation spaces, as well as other miscellaneous background material. Most of the results are summarized in (1.A), and the non-methodical reader may wish to read only this subsection, refering back to the rest as necessary. Section 2 contains the definition of λ (i.e. of $\langle R(W_1), R(W_2) \rangle$). Sections 3 and 4 contain the proof of the Dehn surgery formula. Section 3 contains the parts of the proof involving representation spaces, while Section 4 contains the parts involving the manipulation of surgery diagrams. Some readers may wish to start with Section 4 and refer back to Section 3 as necessary. In Section 5 we use the Dehn surgery formula to give an independent and elementary proof of the existence of λ. This section is independent of sections 1 through 4, and some readers may wish to act accordingly. In Section 6 we use the Dehn surgery formula to prove various things about λ, including its relation to the μ-invariant. Sections A and B are appendices containing background material needed for the statement and proof of the Dehn surgery formula.

This work has previously been announced in [W1]. Certain results found in this paper were obtained independently by S. Boyer and D. Lines [BL] (see (5.1) and (6.5)). A different generalization of Casson's invariant has been described by S. Boyer and A. Nicas [BN].

This research received generous support from the National Science Foundation, the Sloan Foundation and the Mathematical Sciences Research Institute.

It is a pleasure to acknowledge helpful conversations and correspondence with Peter Braam, Dan Freed, William Goldman, Lucien Guillou, Michael Hirsch, Morris Hirsch, Steve Kerckhoff, Gordana Matic, Paul Melvin and Tom Mrowka. Christine Lescop gave an earlier version of this paper a very thorough reading and spotted numerous errors, at least one of which was very non-trivial. Very special thanks are due to Andrew Casson, whose excellent suggestions improved this paper in many ways, and to Rob Kirby, who provided good advice on all sorts of topics. Finally, I would like to thank José Montesinos for suggesting that I work on this problem.

§ 1

Topology of Representation Spaces

A. Summary of Results and Notation.

Let M be an oriented rational homology 3-sphere (QHS) with a genus g Heegaard splitting (W_1, W_2, F). (That is, $M = W_1 \cup W_2$ and $W_1 \cap W_2 = \partial W_1 = \partial W_2 = F$, a surface of genus g.) Let F^* be F minus a disk. We have the following diagram of fundamental groups

$$\begin{array}{ccc} & & \pi_1(W_1) \\ & \nearrow & & \searrow \\ \pi_1(F^*) \to \pi_1(F) & & & & \pi_1(M). \\ & \searrow & & \nearrow \\ & & \pi_1(W_2) \end{array}$$

Note that all maps are surjections. Applying the functor $\mathrm{Hom}(\,\cdot\,, SU(2))$, this diagram becomes

$$\begin{array}{ccc} & & Q_1^\natural \\ & \swarrow & & \searrow \\ R^{*\natural} \leftarrow R^\natural & & & & Q_1^\natural \cap Q_2^\natural \\ & \searrow & & \swarrow \\ & & Q_2^\natural \end{array}$$

(i.e. $R^\natural \stackrel{\text{def}}{=} \mathrm{Hom}(\pi_1(F), SU(2))$, $Q_j^\natural \stackrel{\text{def}}{=} \mathrm{Hom}(\pi_1(W_j), SU(2))$, etc.). Note that all maps are injections and that Van Kampen's theorem implies that $\mathrm{Hom}(\pi_1(M), SU(2)) = Q_1^\natural \cap Q_2^\natural$. $SU(2)$ acts naturally on these spaces via

6

the adjoint action. Taking quotients, we get

$$
\begin{array}{ccc}
 & & Q_1 \\
 & \nearrow & & \searrow \\
R^* \;\leftarrow\; R & & & & Q_1 \cap Q_2 \\
 & \searrow & & \nearrow \\
 & & Q_2
\end{array}
$$

(i.e. $R = \mathrm{Hom}(\pi_1(F), SU(2))/SU(2)$, etc.).

In the rest of this paper the superscript $\sharp$ will be used, without comment, to denote the "inverse quotient by the adjoint action of $SU(2)$", and vice-versa. That is, if X been defined as $Y/SU(2)$, then $X^{\sharp}$ will be used to denote Y.

We will see below that R has two singular strata: S, consisting of (equivalence classes of) representations with abelian image, and P, consisting of representations into $\mathbf{Z}_2$, the center of $SU(2)$. Define

$$
\begin{aligned}
T_j &\overset{\text{def}}{=} Q_j \cap S \\
R^- &\overset{\text{def}}{=} R \setminus S \\
Q_j^- &\overset{\text{def}}{=} Q_j \setminus T_j \\
S^- &\overset{\text{def}}{=} S \setminus P \\
T_j^- &\overset{\text{def}}{=} T_j \setminus P.
\end{aligned}
$$

The Zariski tangent space of $R^{\sharp}$ at a representation ρ can be identified with the cocycle space $Z^1(\pi_1(F); \mathrm{Ad}\,\rho)$, and the tangent space of the orbit through ρ corresponds to the coboundaries $B^1(\pi_1(F); \mathrm{Ad}\,\rho)$. Thus we define the "Zariski tangent space" of R at $[\rho]$ to be $H^1(\pi_1(F); \mathrm{Ad}\,\rho)$.

Let $B : \mathbf{su}(2) \times \mathbf{su}(2) \to \mathbf{R}$ be an Ad-invariant inner product. B induces the cup product

$$
\omega_B : H^1(\pi_1(F); \mathrm{Ad}\,\rho) \times H^1(\pi_1(F); \mathrm{Ad}\,\rho) \to H^1(\pi_1(F); \mathbf{R}) \cong \mathbf{R}.
$$

It is a result of Goldman that ω_B gives R a symplectic structure. Q_1 and Q_2 turn out to be lagrangian with respect to ω_B.

We will be particularly interested in the normal bundle of S^- in R. Define

$$
\begin{aligned}
\nu &\overset{\text{def}}{=} \text{Zariski normal bundle of } S^- \text{ in } R \\
\eta_j &\overset{\text{def}}{=} \text{Zariski normal bundle of } T_j^- \text{ in } Q_j \\
\xi &\overset{\text{def}}{=} \text{actual (singular) normal bundle of } S^- \text{ in } R \\
\theta_j &\overset{\text{def}}{=} \text{actual (singular) normal bundle of } T_j^- \text{ in } Q_j
\end{aligned}
$$

It is another result of Goldman that for $p = [\rho] \in S^-$, the fiber ξ_p is diffeomorphic to a quadratic cone in ν_p modulo $\mathrm{stab}\,(\rho) \cong S^1$. For $p \in T_j^-$, $\theta_{j,p}$ is diffeomorphic to $\eta_{j,p}$ modulo $\mathrm{stab}\,(\rho)$.

Let S_0^1 be a fixed, oriented maximal torus of $SU(2)$ and let $\mathbf{Z}_2 \subset S_0^1$ denote the center of $SU(2)$. Let

$$\widetilde{S} \overset{\mathrm{def}}{=} \mathrm{Hom}(\pi_1(F), S_0^1) \cong (S_0^1)^{2g}$$

$$\widetilde{S}^- \overset{\mathrm{def}}{=} \mathrm{Hom}(\pi_1(F), S_0^1) \setminus \mathrm{Hom}(\pi_1(F), \mathbf{Z}_2) \cong (S_0^1)^{2g} \setminus (\mathbf{Z}_2)^{2g}$$

$$\widetilde{T_j} \overset{\mathrm{def}}{=} \mathrm{Hom}(\pi_1(W_j), S_0^1) \cong (S_0^1)^g$$

$$\widetilde{T_j}^- \overset{\mathrm{def}}{=} \mathrm{Hom}(\pi_1(W_j), S_0^1) \setminus \mathrm{Hom}(\pi_1(W_j), \mathbf{Z}_2) \cong (S_0^1)^g \setminus (\mathbf{Z}_2)^g.$$

$\widetilde{S}^- [\widetilde{T_j}^-]$ is a double cover of $S^- [T_j^-]$. Let ν, η_j, ξ and θ_j continue to denote their lifts to $\widetilde{S}^-$ or $\widetilde{T_j}^-$, as the case may be. Over $\widetilde{S}^-$, ν is a symplectic vector bundle, and η_j is an oriented lagrangian subbundle defined over $\widetilde{T_j}^-$. Picking a metric of F induces an hermetian structure on TR compatible with ω_B. This converts ν into a hermetian vector bundle and η_j into a totally real subbundle.

Let $\det^1(\nu)$ denote the unit vectors in the determinant line bundle $\det(\nu)$ of ν. η_j induces a section $\det^1(\eta_j)$ of $\det^1(\nu)$ over $\widetilde{T_j}^-$. $\det(\nu)\,[\det(\eta_j)]$ extends smoothly and canonically over $\widetilde{S}\,[\widetilde{T_j}]$.

Now we come to a crucial point. Namely, $c_1(\det(\nu)) = c_1(\nu)$ is represented by a multiple ω of ω_B restricted to S^- and lifted to $\widetilde{S}$. Furthermore, $\widetilde{T_j}$ is lagrangian with respect to ω.

The above results are proved (or referenced) in sections B and C. Section D uses results of Newstead to prove that certain maps of representation spaces act trivially on rational cohomology. Section E is concerned with a class of isotopies of R appropriate to the definition of λ. Section F establishes orientation conventions. Section G establishes conventions for Dehn twists and Dehn surgeries.

B. Fundamental Results of Goldman.

The material in this section is treated in more detail in [G1] and [G2].

Let π be a finitely presented group and G be a Lie group. Let $\rho : \pi \to G$ be a representation and let $\rho_t : \pi \to G$ be a differentiable 1-parameter family of representations such that $\rho_0 = \rho$. Writing

$$\rho_t(x) = \exp(tu(x) + O(t^2))\rho(x)$$

(for $x \in \pi$ and t near 0) and differentiating the homomorphism condition

(1.1) $$\rho_t(xy) = \rho_t(x)\rho_t(y),$$

we find that
(1.2) $$u(xy) = u(x) + \operatorname{Ad}\rho(x)\,u(y).$$

In other words, $u : \pi \to \mathbf{g}$ is a 1-cocycle of π with coefficients in the π-module $\mathbf{g}_{\operatorname{Ad}\rho}$. Conversely, solutions of (1.2) lead to maps $\rho_t : \pi \to G$ which satisfy (1.1) to first order in t. Thus the Zariski tangent space of $\operatorname{Hom}(\pi, G)$ at ρ can be identified with $Z^1(\pi; \mathbf{g}_{\operatorname{Ad}\rho})$.

We now compute the tangent space of the Ad-orbit containing ρ. Let g_t be a path in G with $g_0 = 1$. Let

$$\rho_t(x) = g_t^{-1}\rho(x)g_t.$$

If $g_t = \exp(tu_0 + O(t^2))$, then the cocycle corresponding to ρ_t is

$$u(x) = \operatorname{Ad}\rho(x)u_0 - u_0.$$

In other words, u is the coboundary δu_0. Thus the tangent space of the Ad-orbit through ρ is $B^1(\pi; \mathbf{g}_{\operatorname{Ad}\rho})$.

The above results suggest that we define the "Zariski tangent space" of $\operatorname{Hom}(\pi, G)$ at $[\rho]$ to be $H^1(\pi; \mathbf{g}_{\operatorname{Ad}\rho})$. (We will omit the scare quotes from now on.)

We now specialize to the case $\pi = \pi_1(F)$ and G is a Lie group which affords an Ad-invariant, symmetric, non-degenerate bilinear form B on its Lie algebra (e.g. a reductive group with the Cartan-Killing form).

If $X \subset G$, let $Z(X)$ denote the centralizer of X in G. Let $Z(\rho) = Z(\rho(\pi))$.

(1.3) **Proposition (Goldman).** *The dimension of $Z^1(\pi; \mathbf{g}_{\operatorname{Ad}\rho})$ is*

$$(2g-1)\dim G + \dim Z(\rho).$$

The dimension of $H^1(\pi; \mathbf{g}_{\operatorname{Ad}\rho})$ is

$$(2g-2)\dim G + 2\dim Z(\rho).$$

$\square$

Note that $\dim Z^1(\pi; \mathbf{g}_{\mathrm{Ad}\rho})$ is minimal, and hence ρ is a nonsingular point of $\mathrm{Hom}(\pi, G)$, if and only if $\dim(Z(\rho)/Z(G)) = 0$. We denote the set of all such points as $\mathrm{Hom}(\pi, G)^-$. If $\rho \in \mathrm{Hom}(\pi, G)$ is a singular point (i.e. if $\dim(Z(\rho)/Z(G)) > 0$) it is still a nonsingular point of the subvariety $\mathrm{Hom}(\pi, Z(Z(\rho)))^-$. Note also that for all $\sigma \in \mathrm{Hom}(\pi, Z(Z(\rho)))^-$, $\mathrm{stab}\,(\sigma(\pi)) = Z(\sigma) = Z(\rho)$. Therefore all points of $\mathrm{Hom}(\pi, Z(Z(\rho)))^-$ have the same orbit type. In fact,

(1.4) **Proposition (Goldman).** *The sets of the form*

$$\mathrm{Hom}(\pi, Z(Z(X)))^- / N_G(Z(Z(X))),$$

where N_G denotes the normalizer in G, give a stratification of $\mathrm{Hom}(\pi, G)/G$. □

Let $[\rho] \in \mathrm{Hom}(\pi, G)/G$ and $[u] \in H^1(\pi; \mathbf{g}_{\mathrm{Ad}\rho})$. We wish to find necessary and sufficient conditions for $[u]$ to be tangent to a path $[\rho_t]$ in $\mathrm{Hom}(\pi, G)/G$. Writing

$$\rho_t(x) = \exp(tu(x) + t^2 v(x) + O(t^3))\rho(x),$$

plugging this into (1.1), and expanding to second order in t, we find

$$v(x) - v(xy) + \mathrm{Ad}\,\rho(x)v(y) = \frac{1}{2}[u(x), \mathrm{Ad}\,\rho(x)u(y)].$$

The left hand side is just the coboundary of the 1-cochain v. The right hand side is the cup product (on the cocycle level) of u with itself, using the Lie bracket $[\cdot\,, \cdot]: \mathbf{g} \times \mathbf{g} \to \mathbf{g}$ as coefficient pairing. Thus a necessary condition for $[u]$ to be tangent to a path in $\mathrm{Hom}(\pi, G)/G$ is

$$[[u], [u]] = 0 \in H^2(\pi; \mathbf{g}_{\mathrm{Ad}\rho}).$$

It turns out that for surface groups this condition is also sufficient:

(1.5) **Theorem (Goldman).** *Let $[\rho] \in \mathrm{Hom}(\pi, G)/G$ and $\alpha \in H^1(\pi; \mathbf{g}_{\mathrm{Ad}\rho})$. Then α is tangent to a path in $\mathrm{Hom}(\pi, G)/G$ if and only if $[\alpha, \alpha] = 0$.* □

Also,

(1.6) **Theorem (Goldman and Millson, [GM]).** *A point*

$$[\rho] \in \mathrm{Hom}(\pi, G)/G$$

has a neighborhood diffeomorphic to

$$\{\alpha \in H^1(\pi; \mathbf{g}_{\mathrm{Ad}\rho}) \mid [\alpha, \alpha] = 0\}/\mathrm{stab}\,(\rho).$$

□

Recall that we have an Ad-invariant, symmetric, nondegenerate bilinear form $B : \mathbf{g} \times \mathbf{g} \to \mathbf{R}$. This induces a cup product

$$\omega_B : H^1(\pi; \mathbf{g}_{\mathrm{Ad}\rho}) \times H^1(\pi; \mathbf{g}_{\mathrm{Ad}\rho}) \to H^2(\pi; \mathbf{R}) \cong \mathbf{R}.$$

Regarding $H^1(\pi; \mathbf{g}_{\mathrm{Ad}\rho})$ as the Zariski tangent space to $\mathrm{Hom}(\pi, G)/G$ at $[\rho]$, ω_B defines a 2-tensor on $\mathrm{Hom}(\pi, G)/G$.

(1.7) Theorem (Goldman). ω_B *is a closed, nondegenerate exterior 2-form (i.e. a symplectic structure) on* $\mathrm{Hom}(\pi, G)/G$. $\qquad\qquad\square$

ω_B is compatible with the stratification of $\mathrm{Hom}(\pi, G)/G$ in the following sense. Let $[\rho] \in \mathrm{Hom}(\pi, G)/G$ and let $\{\rho_j\} \to \rho$. Then there is a natural inclusion

$$i : \lim H^1(\pi; \mathbf{g}_{\mathrm{Ad}\rho_j}) \to H^1(\pi; \mathbf{g}_{\mathrm{Ad}\rho}),$$

and

$$\lim \omega_{[\rho_j]} = i^* \omega_{[\rho]}.$$

(Here ω_x denotes ω_B restricted to $T_x \mathrm{Hom}(\pi, G)/G$.)

We now apply the above results to the case $G = SU(2)$. Let S_0^1 be a fixed maximal torus of $SU(2)$. Up to conjugacy, $SU(2)$ has three subgroups of the form $Z(Z(X))$: $SU(2)$, S_0^1 and $Z(SU(2)) = \mathbf{Z}_2$. Thus, by (1.4), R $(= \mathrm{Hom}(\pi, SU(2))/SU(2))$ has two singular strata:

$$S \overset{\mathrm{def}}{=} \mathrm{Hom}(\pi, S_0^1)/\mathbf{Z}_2 \cong (S_0^1)^{2g}/\mathbf{Z}_2$$

(here $\mathbf{Z}_2 = N(S_0^1)/Z(S_0^1)$) and

$$P \overset{\mathrm{def}}{=} \mathrm{Hom}(\pi, \mathbf{Z}_2) \cong \mathbf{Z}_2^{2g}.$$

This stratification is compatible with Q_j. That is, Q_j has singular strata $T_j \overset{\mathrm{def}}{=} Q_j \cap S \cong (S_0^1)^g/\mathbf{Z}_2$ and $Q_j \cap P \cong \mathbf{Z}_2^g$.

Let $p = [\rho] \in Q_j$. The inclusion $i : F \to W_j$ induces an injection

$$i^* : H^1(\pi_1(W_j); \mathbf{g}_{\mathrm{Ad}\rho}) \to H^1(\pi; \mathbf{g}_{\mathrm{Ad}\rho})$$

which corresponds to the inclusion $T_p Q_j \to T_p R$. Since $H^2(\pi_1(W_j); \mathbf{R}) = 0$, ω_B is zero on $H^1(\pi_1(W_j); \mathbf{g}_{\mathrm{Ad}\rho})$. Furthermore, $\dim H^1(\pi_1(W_j); \mathbf{g}_{\mathrm{Ad}\rho}) = \frac{1}{2} \dim H^1(\pi; \mathbf{g}_{\mathrm{Ad}\rho})$. Hence Q_j is lagrangian with respect to ω_B.

If $p \in P$, then $\mathrm{Ad}\,\rho$ is trivial, and hence

$$H^1(\pi_1(X); \mathbf{g}_{\mathrm{Ad}\rho}) = H^1(\pi_1(X); \mathbf{R}) \otimes \mathbf{g}$$

for $X = F$ or W_j. Since M is a **QHS**, $H^1(\pi_1(W_1); \mathbf{R})$ and $H^1(\pi_1(W_2); \mathbf{R})$ are transverse in $H^1(\pi_1(F); \mathbf{R})$. Therefore Q_1 and Q_2 are transverse at P.

C. The Normal Bundle of S.

(1.8) The Topology of the Normal Bundle.

Let $p = [\rho] \in S^-$. Let $\mathbf{h}$ be the Lie algebra of S_0^1 and $\mathbf{h}^\perp$ be the orthogonal complement of $\mathbf{h}$ (with respect to B). Then the π-module $\mathbf{g}_{\mathrm{Ad}\rho}$ decomposes as $\mathbf{h}_{\mathrm{Ad}\rho} \oplus \mathbf{h}^\perp_{\mathrm{Ad}\rho}$, and hence

$$H^1(\pi; \mathbf{g}_{\mathrm{Ad}\rho}) = H^1(\pi; \mathbf{h}_{\mathrm{Ad}\rho}) \oplus H^1(\pi; \mathbf{h}^\perp_{\mathrm{Ad}\rho}).$$

The space $H^1(\pi; \mathbf{h}^\perp_{\mathrm{Ad}\rho})$ corresponds to the fiber of the Zariski normal bundle of S^- in R at p. Call this bundle ν. Similarly, if $p \in T_j^-$ also, then $H^1(\pi_1(W_j); \mathbf{h}^\perp_{\mathrm{Ad}\rho})$ is the fiber of the Zariski normal bundle of T_j^- in Q_j at p. Call this bundle η_j.

Let ξ and θ_j denote the actual normal bundles of S^- and T_j^-. By (1.5), we have

$$
\begin{aligned}
\xi &= \{x \in \nu \mid [x, x] = 0\}/S_0^1 \\
\theta_j &= \eta_j/S_0^1.
\end{aligned}
$$

(Here S_0^1 should be thought of as $\mathrm{stab}\,(S_0^1) = \mathrm{stab}\,(\rho(\pi))$.) Let $\hat{\xi}\ [\hat{\theta}_j]$ denote the unit vectors in $\xi\ [\theta_j]$.

It will be convenient to put a hermetian structure on ν compatible with its symplectic structure. This can be done as follows. Let $\mathbf{h}^\perp_{\mathrm{Ad}\rho}$ denote the flat $\mathbf{h}^\perp$-bundle over F with holonomy $\mathrm{Ad}\,\rho$. (The ambiguity in notation is intentional.) Let $H^1_{dR}(F; \mathbf{h}^\perp_{\mathrm{Ad}\rho})$ be the de Rham cohomology group. Since $\pi_i(F) = 0$ for $i \geq 2$, we have $H^1_{dR}(F; \mathbf{h}^\perp_{\mathrm{Ad}\rho}) = H^1(\pi; \mathbf{h}^\perp_{\mathrm{Ad}\rho})$, and we will identify these two spaces from now on. Choose a metric on F. Note that $H^q(F; \mathbf{h}^\perp_{\mathrm{Ad}\rho})$ can also be identified with the space of harmonic q-forms with coefficients in $\mathbf{h}^\perp_{\mathrm{Ad}\rho}$.

Choose B to be positive definite (e.g. the negative of the Cartan-Killing form). The metric on F, together with B, induces the Hodge star operator

$$* : H^1(\pi; \mathbf{h}^\perp_{\mathrm{Ad}\rho}) \to H^1(\pi; \mathbf{h}^\perp_{\mathrm{Ad}\rho})$$

and the Hodge metric

$$\langle \cdot, \cdot \rangle : H^1(\pi; \mathbf{h}^\perp_{\mathrm{Ad}\rho}) \times H^1(\pi; \mathbf{h}^\perp_{\mathrm{Ad}\rho}) \to \mathbf{R},$$

where

$$\langle \alpha, \beta \rangle \overset{\text{def}}{=} \int_F B(\alpha, *\beta).$$

It is easy to see that $*$ and $\langle \cdot, \cdot \rangle$ give a hermetian structure compatible with ω_B. That is,

$$\omega_B(\alpha, \beta) = -\langle \alpha, *\beta \rangle$$

for all $\alpha, \beta \in H^1(\pi; \mathbf{h}_{\text{Ad}\rho}^\perp)$.

$H^1(\pi; \mathbf{h}_{\text{Ad}\rho}^\perp)$ has a second complex structure arising from the coefficient module $\mathbf{h}_{\text{Ad}\rho}^\perp$. Fix a positively oriented vector $v \in \mathbf{h}$. After possibly rescaling B, there is a map $J : \mathbf{h}^\perp \to \mathbf{h}^\perp$ such that for all $a, b \in \mathbf{h}^\perp$,

(1.9) $$B([a, b], v) = B(a, Jb).$$

The action of $\text{Ad}(S_0^1)$ commutes with J, and $J^2 = -1$. Thus $H^1(\pi; \mathbf{h}_{\text{Ad}\rho}^\perp)$ inherits a complex structure from $\mathbf{h}^\perp$, also denoted J. The action of $\text{Ad}(S_0^1)$ can be viewed as multiplication by unit complex numbers (with weight 2). However, in order for J to give a complex structure on the *bundle* ν, it must be lifted to $\widetilde{S}^-$, since the action of $N(S_0^1)/Z(S_0^1)$ is *conjugate* linear. For the remainder of this section, regard ν, η_j, etc. as so lifted.

Note that J and $*$ commute. Note also that while η_j is totally real with respect to $*$, it is complex with respect to J. Call such subspaces complex lagrangians.

It follows from (1.9) that

$$[\alpha, \beta] = \langle \alpha, J * \beta \rangle$$

for all $\alpha, \beta \in \nu_p$. Since $(J*)^2 = 1$, we can decompose ν as $A^+ \oplus A^-$, where $J * |_{A^\pm} = \pm 1$. Since η_j is totally real with respect to $*$ and complex with respect to J, we must have $\dim A^\pm = \frac{1}{2} \dim \nu = 2g - 2$. Note that $* = -J$ on A^+ and $* = J$ on A^-.

The solutions to $[\alpha, \alpha] = 0$ are the cone on the product of the unit sphere bundles of A^+ and A^-. Thus

$$\hat{\xi}_p \cong X \overset{\text{def}}{=} (S^{2g-3} \times S^{2g-3})/S^1,$$

where S^1 acts diagonally via complex multiplication on $S^{2g-3} \subset \mathbf{C}^{g-1}$. Also,

$$\hat{\theta}_{j,p} \cong \mathbf{C}P^{g-2}.$$

We wish to analyze $\mathcal{L}_p^C$, the space of all complex lagrangians of ν_p. Consider a complex lagrangian $L \in \mathcal{L}_p^C$. L must be transverse to A^+ and A^-, and so can be thought of as a graph of a linear map $T : A^+ \to A^-$.

Since L is complex with respect to J, T must be J-linear (and hence *-conjugate-linear). Since L is totally real with respect to *, we must have, for all $\alpha, \beta \in A^+$,

$$
\begin{aligned}
0 &= \langle (\alpha, T\alpha), *(\beta, T\beta) \rangle \\
&= \langle \alpha, *\beta \rangle + \langle T\alpha, *T\beta \rangle \\
&= -\langle \alpha, J\beta \rangle + \langle T\alpha, TJ\beta \rangle.
\end{aligned}
$$

Therefore T is unitary with respect to J, and $\mathcal{L}_p^C$ can be identified with the set $U(A_p^+, A_p^-)$ of all J-unitary maps from A_p^+ to A_p^-.

(1.10) Two Homomorphisms.

The homomorphisms defined in the next two paragraphs will be used in Section 2.

Let $\det(\nu_p) = \bigwedge^{2g-2} \nu_p$, where here ν_p is viewed as a *-complex vector space. Let $\det^1(\nu_p) = \det(\nu_p)/\mathbf{R}^+ \cong S^1$. There is a map

$$(1.11) \qquad \det^1 : \mathcal{L}_p^C \to \det^1(\nu_p)$$

defined as follows. Elements of $\det(\nu_p)$ are represented by bases of ν_p (over $\mathbf{C}$), and two such bases are identified if the linear map connecting them has determinant 1. Given $L \in \mathcal{L}_p^C$, choose a basis (over $\mathbf{R}$) of L which is oriented with respect to the orientation coming from the J-complex structure. This will also be a basis of ν_p over $\mathbf{C}$, with respect to *. Since L is totally real with respect to *, any two such bases differ by an element of $GL^+(2g - 2, \mathbf{R}) \subset GL(2g - 2, \mathbf{C})$ (well-defined up to conjugacy). Since $\det(GL^+(2g - 2, \mathbf{R})) = \mathbf{R}^+$, this basis represents a well-defined element of $\det^1(\nu_p)$. Now define

$$
\begin{aligned}
\varphi : \quad \pi_1(\mathcal{L}_p^C) &\to \mathbf{Z} \\
[\alpha] &\mapsto \deg(\det^1(\alpha)),
\end{aligned}
$$

where $\det^1(\alpha)$ is viewed as a map from S^1 to S^1, and deg denotes the degree of the map. φ is clearly a homomorphism.

Now we define another homomorphism

$$\psi : \pi_1(\mathcal{L}_p^C) \to \mathbf{Z}.$$

By taking unit vectors and dividing out by the action of $\mathrm{Ad}\,(S_0^1)$, each $L \in \mathcal{L}_p^C$ gives rise to a complex projective space $\mathbf{C}P(L) \subset \hat{\xi}_p$. (e.g. $\mathbf{C}P(\eta_{j,p}) = \hat{\theta}_{j,p}$.) Orient $\mathbf{C}P(L)$ according to its J-complex structure. Orient $\hat{\xi}_p$ so that its orientation followed by an outward pointing radial vector gives the symplectic orientation of ξ_p. Choose $L_0 \in \mathcal{L}_p^C$. Let $[\alpha] \in \pi_1(\mathcal{L}_p^C)$. Define

$$\psi([\alpha]) \overset{\text{def}}{=} \langle CP(\alpha), CP(L_0) \rangle_{\hat{\xi}_p}.$$

(Here $\langle \, \cdot \, , \, \cdot \, \rangle_{\hat{\xi}_p}$ denotes the intersection number in $\hat{\xi}_p$.) It is easy to see that $\psi([\alpha])$ depends only on the homotopy class of α and is independent of the choice of L_0.

(1.12) **Lemma.** $\varphi = (-1)^{g-1}\psi$.

Proof: Since $\pi_1(\mathcal{L}_p^C) \cong \pi_1(U(g-1)) \cong \mathbf{Z}$, it suffices to compare φ and ψ on a non-zero element of $\pi_1(\mathcal{L}_p^C)$.

Fix $T \in U(A_p^+, A_p^-) = \mathcal{L}_p^C$. Let U^- be the unitary transformations of A_p^-. Identify $V \in U^-$ with $VT \in U(A_p^+, A_p^-)$. Let $\alpha : S^1 \to U^-$ be the diagonal embedding of S^1. (The orientation is unambiguous, since $J = *$ on A_p^-.) It is easy to verify that

$$\varphi([\alpha]) = g - 1.$$

Now we compute

$$\psi([\alpha]) = \langle CP(\alpha), CP(T) \rangle_{\hat{\xi}_p}.$$

Since $CP(T) \subset CP(\alpha)$, the above intersection number is equal to the self intersection of $CP(T)$ inside the normal bundle of $CP(\alpha)$ restricted to $CP(T)$. To get the sign right, the orientation of this restricted bundle should be such that it gives the orientation of $\hat{\xi}_p$ when followed by the standard orientation of the "S^1 direction" in $CP(\alpha)$. This self intersection number is the same as the self intersection of the image of $CP(T)$ in $\hat{\xi}_p / \alpha(S^1)$. ($\alpha(S^1)$ is a subgroup of U^-, and it acts on $\hat{\xi}_p$ in the obvious way.) It is not hard to see that the pair $(\hat{\xi}_p / \alpha(S^1), CP(T))$ is diffeomorphic to $(CP^{g-2} \times CP^{g-2}, \Delta)$, where Δ is the diagonal. $\hat{\xi}_p / \alpha(S^1)$ can also be identified with $CP(A_p^+) \times CP(A_p^-)$. $CP(A_p^+) \times CP(A_p^-)$ has three orientations of interest: O_c, the orientation which makes the sign of the self intersection come out correctly; O_J, its orientation as a J-complex manifold; and O_*, its orientation as a $*$-complex manifold. It turns out that $O_c = -O_*$. Also, since $J = -*$ on A_p^+ and $\dim_C(CP(A_p^+)) = g - 2$, $O_* = (-1)^{g-2}O_J$. Since Δ is a J-complex manifold, the self intersection of Δ in $CP(A_p^+) \times CP(A_p^-)$ with respect to O_J is equal to $g - 1$, the Euler characteristic of CP^{g-2}. Hence

$$\psi([\alpha]) = (-1)^{g-1}(g - 1).$$

□

(1.13) Extending $\det(\nu)$.

Consider the bundle $\det(\nu)$ over $\widetilde{S}^-$. Our next goal is to show that $\det(\nu)$ extends naturally over all of $\widetilde{S}$ and that the fiber over $\rho \in P$ is naturally identified with $\det(H^1(F; \mathbf{h}_{\mathrm{Ad}\rho}^\perp))$. Furthermore, η_j gives rise to a section $\det^1(\eta_j)$ of $\det^1(\nu)$ over $\widetilde{T_j}^-$ which extends continuously to $\det^1(H^1(W_j; \mathbf{h}_{\mathrm{Ad}\rho}^\perp))$ over $\rho \in \widetilde{T_j} \cap P$. To prove these facts we view $\det(\nu)$ as the determinant line bundle of a family of Fredholm operators (cf. [Q]).

Let $\Omega^q(\mathbf{h}^\perp)$ be the space of q-forms on F with values in $\mathbf{h}^\perp$. The Hodge star operator gives a complex structure on $\Omega^1(\mathbf{h}^\perp)$ and $\Omega^0(\mathbf{h}^\perp) \oplus \Omega^2(\mathbf{h}^\perp)$. Let E be the trivial principal S_0^1 bundle over F. Let $\mathcal{F}$ be the space of flat orthogonal connections on E. $\mathcal{F}$ can also be viewed, via the representation $\mathrm{Ad}(S_0^1)$, as the space of flat connections on $F \times \mathbf{h}^\perp$. Let $\mathcal{G}_0$ be the gauge transformations of E which are the identity at some fixed point. $\mathcal{G}_0$ acts freely on $\mathcal{F}$ and $\mathcal{F}/\mathcal{G}_0$ can be identified with $\widetilde{S}$.

For each $A \in \mathcal{F}$, we have the exterior covariant derivative

$$d_A : \Omega^q(\mathbf{h}^\perp) \to \Omega^{q+1}(\mathbf{h}^\perp)$$

and its adjoint

$$d_A^* = - * d_A * : \Omega^{q+1}(\mathbf{h}^\perp) \to \Omega^q(\mathbf{h}^\perp).$$

Let D_A be the operator

$$D_A = d_A + d_A^* : \Omega^1(\mathbf{h}^\perp) \to \Omega^0(\mathbf{h}^\perp) \oplus \Omega^2(\mathbf{h}^\perp).$$

The kernel of D_A consists of the harmonic 1-forms (with respect to the flat structure given by A), which can be identified with $H^1(F; \mathbf{h}_{\mathrm{Ad}\rho}^\perp)$, where ρ is the holonomy representation of A. Similarly, $\mathrm{coker}\, D_A = H^0(F; \mathbf{h}_{\mathrm{Ad}\rho}^\perp) \oplus H^2(F; \mathbf{h}_{\mathrm{Ad}\rho}^\perp)$.

Following Quillen ([Q]), we now describe a bundle $\det(D)$ over $\mathcal{F}$, whose fiber at A can be identified with $\det(\ker D_A) \otimes \det(\mathrm{coker}\, D_A)^*$. (Note that this is non-trivial, since the dimensions of $\ker D_A$ and $\mathrm{coker}\, D_A$ are not constant.) Let V be any finite dimensional subspace of $\Omega^0(\mathbf{h}^\perp) \oplus \Omega^2(\mathbf{h}^\perp)$. Let U_V be the set of all $A \in \mathcal{F}$ such that D_A is transverse to V. (That is, $\mathrm{im}(D_A) + V = \Omega^0(\mathbf{h}^\perp) \oplus \Omega^2(\mathbf{h}^\perp)$.) For such A there is an exact sequence

$$0 \to \ker D_A \to D_A^{-1}(V) \to V \to \mathrm{coker}\, D_A \to 0$$

which induces a canonical isomorphism

$$(1.14) \quad \det(\ker D_A) \otimes \det(\operatorname{coker} D_A)^* \cong \det(D_A^{-1}(V)) \otimes \det(V)^*.$$

The set U_V is open, and $D_A^{-1}(V)$ is a smooth vector bundle over U_V. Thus the right hand side of (1.14) is a smooth complex line bundle over U_V. Such bundles over the sets U_V (for various V) fit together to form $\det(D)$ over all of $\mathcal{F}$.

$\mathcal{G}_0$ acts equivariantly on $\det(D)$, yielding a bundle over $\mathcal{F}/\mathcal{G}_0 = \widetilde{S}$, which we also denote by $\det(D)$. If $\rho \in \widetilde{S}^-$, then $H^0(F; \mathbf{h}_{\operatorname{Ad}\rho}^\perp) \cong H^2(F; \mathbf{h}_{\operatorname{Ad}\rho}^\perp) \cong 0$. Hence $\det(D)_\rho = \det(H^1(F; \mathbf{h}_{\operatorname{Ad}\rho}^\perp)) = \det(\nu_\rho)$. If $\rho \in P$, then $H^0(F; \mathbf{h}_{\operatorname{Ad}\rho}^\perp)$ is a complex lagrangian in $H^0(F; \mathbf{h}_{\operatorname{Ad}\rho}^\perp) \oplus H^2(F; \mathbf{h}_{\operatorname{Ad}\rho}^\perp) = \operatorname{coker} D_\rho$. This gives a canonical element of $\det(\operatorname{coker} D_\rho)$, which allows us to identify $\det(D)_\rho$ with $\det(\ker D_\rho) = \det(H^1(F; \mathbf{h}_{\operatorname{Ad}\rho}^\perp))$. So $\det(D)$ is the desired extension of $\det(\nu)$.

The map $\det^1 : \mathcal{L}^C \to \det^1(\nu)$ of (1.11), applied to η_j, gives a section $\det^1(\eta_j)$ of $\det^1(\nu)$ defined over $\widetilde{T_j}^-$. We wish to show that this section extends smoothly over all of $\widetilde{T_j}$.

Fix a trivial connection $A_0 \in \mathcal{F}$. Let $d = d_{A_0}$. Let H^q be the the harmonic q-forms with coefficients in $\mathbf{h}^\perp$ (with respect to A_0). Let $H^1(F; \mathbf{h})$ be the harmonic 1-forms with coefficients in $\mathbf{h}$. Given $\rho \in \widetilde{S}$ near 1, there is a unique $\alpha_\rho \in H^1(F; \mathbf{h})$ near zero such that for all loops $\gamma \subset F$,

$$\exp\left(\int_\gamma \alpha_\rho \right) = \rho(\gamma).$$

Let

$$d_\rho = d + \alpha_\rho.$$

The map $\rho \mapsto d_\rho$ is a section of the fibration $\mathcal{F} \to \widetilde{S}$, for ρ near 1. Let

$$D_\rho = d_\rho + d_\rho^*,$$

and let H_ρ^q be the harmonic q-forms with values in $\mathbf{h}^\perp$ with respect to d_ρ. That is, $H_\rho^q = \ker(D_\rho)$. Note that for $q = 0$ or 2, $H_\rho^q = H^q$ for $\rho = 1$ and $H_\rho^q = 0$ for $\rho \neq 1$.

$D_\rho : \Omega^1(\mathbf{h}^\perp) \to \Omega^0(\mathbf{h}^\perp) \oplus \Omega^2(\mathbf{h}^\perp)$ is transverse to $H^0 \oplus H^2$ for ρ near 1. Hence $\det(\nu)$ can be identified with the determinant of the (finite dimensional) operator

$$D_\rho : D_\rho^{-1}(H^0 \oplus H^2) \to H^0 \oplus H^2$$

for ρ near 1.

Consider the diagram

$$
\begin{array}{ccc}
D_\rho^{-1}(H^0 \oplus H^2) & \overset{D_\rho}{\to} & H^0 \oplus H^2 \\
\downarrow \pi & & \downarrow i \\
H^1 & \overset{D_\rho'}{\to} & \mathbf{h}^\perp \oplus \mathbf{h}^\perp ,
\end{array}
$$

where π is orthogonal projection (with respect to the Hodge inner product), i is the isomorphism

$$
i(a, b) \overset{\text{def}}{=} (\int_F *a, \int_F b),
$$

and

$$
D_\rho'(a) \overset{\text{def}}{=} (\int_F \alpha_\rho \wedge *a, \int_F \alpha_\rho \wedge a).
$$

Using the Hodge decomposition theorem, it is not hard to see that this diagram commutes, and that π is an isomorphism for ρ near 1. Hence, for such ρ, $\det(\nu)$ can be identified with the determinant of D_ρ'.

The lagrangian $H^1(W_1; \mathbf{h}^\perp) \otimes (\mathbf{h}^\perp \oplus 0)^* \subset H^1 \otimes (\mathbf{h}^\perp \oplus \mathbf{h}^\perp)^*$ gives rise to a smooth section of $\det^1(D_\rho') = \det^1(\nu)$ near $\rho = 1$. For $\rho \neq 1$, this section coincides with $\det^1(\eta_1)$. For $\rho = 1$, it coincides with $\det^1(H^1(W_1; \mathbf{h}^\perp))$. Similar things are true if W_1 is replaced by W_2 or if 1 is replaced with another representation in P. Thus we have found the desired extensions of $\det^1(\nu)$ and $\det^1(\eta_j)$.

As a notational contrivance, let us define, for $\rho \in P \subset \tilde{S}$,

$$
\nu_\rho \overset{\text{def}}{=} H^1(F; \mathbf{h}_{\mathrm{Ad}\rho}^\perp) \cong H^1(F; \mathbf{h}^\perp).
$$

If $\rho \in P \cap \widetilde{T_j}$, define

$$
\eta_{j,\rho} \overset{\text{def}}{=} H^1(W_j; \mathbf{h}_{\mathrm{Ad}\rho}^\perp) \cong H^1(W_j; \mathbf{h}^\perp).
$$

So, abusing notation slightly, $\det^1(\nu)$ is a smooth bundle over all of $\tilde{S}$ and $\det^1(\eta_j)$ is a smooth section defined over all of $\widetilde{T_j}$. This will simplify notation in what follows.

(1.15) **Proposition.** *The first Chern class $c_1(\det(\nu))$ of $\det(\nu)$ is represented by a multiple ω of ω_B. If $T^2 \subset \tilde{S}$ is a symplectic 2-torus corresponding to a 2-dimensional symplectic summand of $H_1(F; \mathbf{R})$, then $\int_{T^2} \omega = -8$.*

Proof: Let $\mathcal{A}$ be the space of connections on $F \times S_0^1$. The first step is to compute the curvature the universal S_0^1 bundle $\mathcal{U} = \mathcal{A} \times F \times S_0^1$ over $\mathcal{A} \times F$ (cf. [AS]).

Identify $\mathcal{A}$ with $\Omega^1(F; \mathbf{h})$ via $A \mapsto d + A$. Identify $T_A\mathcal{A}$ with $\Omega^1(F; \mathbf{h})$. Let $(A, x, \varphi) \in \mathcal{U}$ and $(a, X, w) \in T_{(A,x,\varphi)}\mathcal{U} = \Omega^1(F; \mathbf{h}) \times T_x F \times \mathbf{h}$. Define the connection 1-form θ by

$$\theta_{(A,x,\varphi)}(a, X, w) = A(X) + w.$$

The connection θ has the property that over the surface $\{A\} \times F$ it restricts to the connection A on $F \times S_0^1$. The curvature of θ is $\Theta = d\theta \in \Omega^2(\mathcal{A} \times F; \mathbf{h})$. Thus, for $(a, X), (b, Y) \in T_{(A,x)}\mathcal{A} \times F$,

$$\Theta((a, X), (b, Y)) = a(Y) - b(X) + dA(X, Y).$$

Let $\mathcal{G}_0$ be the group of smooth maps $g : F \to S_0^1$ such that $g(x_0) = 1$ for some fixed $x_0 \in F$. $\mathcal{G}_0$ acts freely on $\mathcal{U}$. This action preserves the connection θ, and so leads to a connection θ^b on $\mathcal{U}^b \overset{\text{def}}{=} \mathcal{U}/\mathcal{G}_0$. Recall that $\widetilde{S}$ can be identified with $\mathcal{F}/\mathcal{G}_0$, where $\mathcal{F} \subset \mathcal{A}$ is the space of flat connection on $F \times S_0^1$. Thus θ^b restricts to a connection on $\widetilde{S} \times F$.

The representation $S_0^1 \to \text{Ad}\,(S_0^1)|_{\mathbf{h}^\perp}$ associates to $\mathcal{U}^b$ a $\mathbf{h}^\perp$ bundle E over $\widetilde{S} \times F$ with curvature

$$\Theta' = \begin{bmatrix} 0 & 2\Theta \\ -2\Theta & 0 \end{bmatrix}.$$

Hence $c_1(E \otimes \mathbf{C}) = (2\pi i)^{-1}\text{tr}(\Theta') = 0$ and $c_2(E \otimes \mathbf{C})$ is represented by

$$-\frac{1}{4\pi^2}\det(\Theta') = -\frac{1}{\pi^2}\Theta \wedge \Theta.$$

Thus the Chern character of E is represented by

$$\text{ch}(E) = 2 + \frac{1}{\pi^2}\Theta \wedge \Theta.$$

E can be thought of as a family of flat $\mathbf{h}^\perp$ bundles over F parameterized by $\widetilde{S}$. Corresponding to each flat connection A is the operator D_A, defined above, and $\det(\nu)$ can be identified with the determinant of the index bundle of this family.

A flat connection A on $F \times (\mathbf{h}^\perp \otimes \mathbf{C})$ determines a holomorphic structure A' on $F \times (\mathbf{h}^\perp \otimes \mathbf{C})$, and the index of the family D_A can be identified with the index of the family $\bar{\partial}_{A'}$. (F is given the complex structure compatible with its metric.) So, by Theorem 5.1 of [AS4] and the Riemann-Roch theorem, $c_1(\det(\nu))$ is represented by the 2-dimensional part of

$$\int_F \text{ch}(E)T(F),$$

where $\int_F$ denotes integration along the fibers (i.e. along F) and $T(F)$ is a form representing the Todd genus of F. Since $T(F) = 1 + \alpha$, where $\alpha \in \Omega^2(F)$, this is just

$$\int_F \frac{1}{\pi^2}\Theta \wedge \Theta.$$

Note that, for $a, b \in H^1(F; \mathbf{h}) \cong T_p\tilde{S}$ and $X, Y \in T_x F$,

$$\begin{aligned}\Theta \wedge \Theta(a, b, X, Y) &= -a(X)b(Y) + a(Y)b(X) + b(X)a(Y) - b(Y)a(X) \\ &= -2a \wedge b(X, Y).\end{aligned}$$

Hence $c_1(\det(\nu))$ is represented by the 2-form ω, where

$$\omega(a, b) \overset{\text{def}}{=} -\frac{2}{\pi^2}a \wedge b.$$

(Here we have identified $H^2(F; \mathbf{R})$ with $\mathbf{R}$ via $c \mapsto \int_F c$.) Let $T^2 \subset \tilde{S}$ be a symplectic 2-torus. Then

$$\int_{T^2} \omega = -\frac{2}{\pi^2}(2\pi)^2 = -8.$$

$\square$

D. Results of Newstead.

In this subsection we use results of Newstead ([N1,N2]) to show that certain diffeomorphisms of representation spaces are homologically trivial.

Let $\gamma \subset F$ be a separating simple closed curve. Define

$$N_\gamma \overset{\text{def}}{=} \{[\rho] \in R \,|\, \rho(\gamma) = -1\}.$$

Let F_1 and F_2 be the components of F cut along γ. Define

$$R_j^{*\natural} \overset{\text{def}}{=} \text{Hom}(\pi_1(F_j), SU(2))$$
$$N_j^\natural \overset{\text{def}}{=} \{\rho \in R_j^{*\natural} \,|\, \rho(\gamma) = -1\}$$

$(j = 1, 2)$. Then we can make the identifications

$$
\begin{aligned}
R^{\sharp} &= \{(\rho_1, \rho_2) \in R_1^{*\sharp} \times R_2^{*\sharp} \mid \rho_1(\gamma) = \rho_2(\gamma)\} \\
N_\gamma^{\sharp} &= N_1^{\sharp} \times N_2^{\sharp}.
\end{aligned}
$$

(In the first equation, some basing of the free loop γ is assumed.)

Let $\tau : F_1 \to F_1$ be a diffeomorphism such that $\tau|_\gamma = \mathrm{id}$ and $\tau_* : H_*(F_1; \mathbf{Z}) \to H_*(F_1; \mathbf{Z})$ is the identity. τ induces diffeomorphisms of F (which is the identity on F_2) and N_γ, which will also be denoted τ. In the proofs of (3.10) and (3.12), we will need the following lemma.

(1.16) **Lemma.** $\tau^* : H^{3g-3}(N_\gamma; \mathbf{Q}) \to H^{3g-3}(N_\gamma; \mathbf{Q})$ *is the identity.*

Proof: Note that since $[\gamma]$ lies in the commutator subgroup of $\pi_1(F)$, $N_\gamma^{\sharp} = N_1^{\sharp} \times N_2^{\sharp}$ consists entirely of irreducible representations. Hence there is a fibration

(1.17) $$ p : N_1^{\sharp} \times N_2^{\sharp} \to N_\gamma $$

with fiber $SU(2)/Z(SU(2)) \cong SO(3)$. Since $SO(3)$ is a rational homology 3-sphere, there is a Gysin sequence

$$ \cdots \to H^{i-4}(N_\gamma) \xrightarrow{\cup \chi} H^i(N_\gamma) \xrightarrow{p^*} H^i(N_1^{\sharp} \times N_2^{\sharp}) \to H^{i-3}(N_\gamma) \to \cdots . $$

(All coefficients are in $\mathbf{Q}$. $\chi \in H^4(N_\gamma)$ is the Euler class of the fibration.)

I claim that $p^* : H^i(N_\gamma) \to H^i(N_1^{\sharp} \times N_2^{\sharp})$ is an isomorphism for $i = 2$ or 3. This is obvious for $i = 2$, and injectivity is obvious for $i = 3$. By Proposition 2.6 of [N2], the natural map

$$ H^3(N_j) \to H^3(N_j^{\sharp}) $$

is an isomorphism $(j = 1, 2)$. Furthermore, $H^1(N_j) \cong H^1(N_j^{\sharp}) \cong 0$ (Theorems 1 and 1$'$ of [N2]). Therefore, by the Künneth formula, the composite

$$ H^3(N_1 \times N_2) \to H^3(N_\gamma) \to H^3(N_1^{\sharp} \times N_2^{\sharp}) $$

is an isomorphism. (The first arrow is induced by the fibration $N_\gamma \to N_1 \times N_2$.) Hence $p^* : H^3(N_\gamma) \to H^3(N_1^{\sharp} \times N_2^{\sharp})$ is onto.

By Theorem 1$'$ of [N2] and the Künneth formula, $H^i(N_1^{\sharp} \times N_2^{\sharp})$ is generated by classes of dimensions 2 and 3 for $i \leq 3g - 3$. I claim that $H^i(N_\gamma)$ is generated by classes of dimensions 2 and 3 and $\chi \in H^4(N_\gamma)$, for $i \leq 3g - 3$. This is clear for $i \leq 3$. Suppose it holds for $i \leq k < 3g - 3$. Let $\alpha \in H^{k+1}(N_\gamma)$. Then there exists $\beta \in H^{k+1}(N_\gamma)$ lying in the subring of

$H^*(N_\gamma)$ generated by classes of dimensions 2 and 3 such that $p^*(\alpha - \beta) = 0$. Hence

$$\alpha = \beta + \eta \cup \chi,$$

where, by inductive assumption, $\eta \in H^{k-3}(N_\gamma)$ lies in the subring of $H^*(N_\gamma)$ generated by classes of dimensions 2 and 3 and χ.

So it suffices to show that τ^* is the identity on $H^2(N_\gamma)$, $H^3(N_\gamma)$, and χ.

Since τ acts on the fibration (1.17) in an orientation preserving fashion, $\tau^*(\chi) = \chi$.

Since $p^* : H^i(N_\gamma) \to H^i(N_1^\sharp \times N_2^\sharp)$ is an isomorphism for $i = 2$ or 3, it suffices to show that τ acts trivially on $H^i(N_1^\sharp \times N_2^\sharp)$. By the Künneth formula and the fact that τ act trivially on $H^i(N_2^\sharp)$, it suffices to show that τ acts trivially on $H^i(N_1^\sharp)$ ($i = 2, 3$). This is done in [AM] (Lemmas VI.2.1 and VI.2.2). □

Let $x_1, y_1, \ldots, x_k, y_k$ be a symplectic basis of $\pi_1(F_1)$. This gives rise to identifications

$$R_1^{*\sharp} = (SU(2))^{2k}$$

$$N_1^\sharp = \{(X_1, Y_1, \ldots, X_k, Y_k) \in R_1^{*\sharp} \mid \prod_1^k [X_i, Y_i] = -1\}.$$

Define $\sigma : N_1^\sharp \to N_1^\sharp$ by

$$\sigma(X_1, Y_1, \ldots, X_k, Y_k) = (X_1, -Y_1, X_2, Y_2, \ldots, X_k, Y_k).$$

$\sigma \times \mathrm{id} : N_1^\sharp \times N_2^\sharp$ descends to a diffeomorphism of N_γ, which will also be denoted by σ. In the proof of (3.59), we will need the following lemma.

(1.18) **Lemma.** $\sigma^* : H^{3g-3}(N_\gamma; \mathbf{Q}) \to H^{3g-3}(N_\gamma; \mathbf{Q})$ *is the identity.*

Proof: The proof of (1.16) can be adapted to this case almost verbatim. It is left to the reader to check that the proofs of Lemmas VI.2.1 and VI.2.2 of [AM] work with τ replaced by σ. (This amounts to observing that $\sigma^* : H^3(R_1^{*\sharp}) \to H^3(R_1^{*\sharp})$ is the identity and that σ acts on various sphere bundles in an orientation preserving fashion.) □

E. Special Isotopies.

(1.19) *Definition.* An isotopy $\{h_t\}_{0 \le t \le 1}$ of R is called *special* if

1. $h_t(Q_1)$ is transverse to Q_2 at $Q_1 \cap Q_2 \cap P$ for all t.

2. $h_t|_S = \mathrm{id}$ for all t.

3. $(h_t)_* : TR|_{S^-} \to TR|_{S^-}$ is symplectic, and hence (in view of 2 above) preserves the fibers of the normal bundle ν.

(1.20) **Proposition.** *There exists a special isotopy $\{h_t\}_{0 \leq t \leq 1}$ of R such that $h_1(Q_1)$ is transverse to Q_2 (i.e. their Zariski tangent spaces are transverse at each point of $Q_1 \cap Q_2$).*

Proof: Since M is a **QHS**, Q_1 is transverse to Q_2 at P and T_1 is transverse to T_2 in S. Choose a compactly supported isotopy of a tubular neighborhood of S^- in R which moves η_{1p} transverse to η_{2p} for each $p \in T_1^- \cap T_2^-$ and is symplectic on $TR|_{S_-}$. At this stage Q_1 is transverse to Q_2 in a neighborhood of S, and so we can find a compactly supported isotopy of R^- which moves Q_1 transverse to Q_2. $\qquad\qquad\qquad\qquad \Box$

F. Orientations.

Orientations will be important in what follows, so in this section we will establish orientation conventions. $[Y]$ will denote the orientation of the space Y. If Y is singular, this means the orientation of the top stratum. If Y is a bundle, it means an orientation of the fibers (not the total space).

First of all, orient the Heegaard surface F so that $[F]$ followed by a normal vector to F pointing into Q_2 gives $[M]$. This fixes an identification of $H^2(\pi_1(F); \mathbf{R})$ with $\mathbf{R}$, and so fixes the sign of ω.

Complex vector spaces (and almost complex manifolds) have a natural orientation: If $a_1, \ldots, a_n$ is a basis over $\mathbf{C}$, then $a_1, ia_1, \ldots, a_n, ia_n$ is an oriented basis over $\mathbf{R}$. Symplectic vector spaces and manifolds are oriented according to a compatible almost complex structure. Thus ω determines the orientations $[R]$, $[S]$, $[\nu]$ and $[\xi]$. η_j and $\hat{\theta}_j$, when lifted to $\widetilde{T_j}^-$, have J-complex structures. This determines $[\eta_j]$ and $[\hat{\theta}_j]$ (as bundles over $\widetilde{T_j}^-$).

Choose orientations of $\widetilde{T_1}$ and $\widetilde{T_2}$ so that

$$[\widetilde{S}] = [\widetilde{T_1}][\widetilde{T_2}]$$

at points of $\widetilde{T_1} \cap \widetilde{T_2}$. ($[\widetilde{S}]$ is the orientation lifted from $[S]$.)

In general, given a fibering $Y \hookrightarrow E \to B$, we choose orientations so that $[E] = [B][Y]$. Thus orientations on two of E, B or Y determine an orientation of the third. If G acts on X, there is a fibering $G \hookrightarrow$

$X \rightarrow X/G$ (at least at points where G acts freely, which is enough to determine orientations). We regard spaces of unit vectors (e.g. $\hat{\theta}_j$) as quotients of spaces of non-unit vectors (e.g. θ_j) by $\mathbf{R}^+$, the positive reals. Let $[SU(2)]$, $[S_0^1]$ and $[\mathbf{R}^+]$ be the standard orientations. The following equations determine orientations of spaces not yet oriented. (There are some cases of over determination, and it is left to the reader to check that these cases are consistent.)

$$
\begin{aligned}
[R] &= [S][\xi] \\
[\xi] &= [\hat{\xi}][\mathbf{R}^+] \\
[Q_j] &= [\widetilde{T_j}][\theta_j] \\
[\eta_j] &= [\theta_j][S_0^1] \\
[\theta_j] &= [\hat{\theta}_j][\mathbf{R}^+] \\
[Q_j^\natural] &= [Q_j][SU(2)] \\
[R^\natural] &= [R][SU(2)]
\end{aligned}
$$

(The third equation requires some comment. $[\widetilde{T_j}][\theta_j]$ determines an orientation of a double cover of a neighborhood of T_j^- in Q_j, which induces an orientation on Q_j.)

All boundaries will be oriented according to the "inward normal last" convention. (e.g. F is oriented as ∂W_2.) In particular, this fixes an orientation of ∂F^*. Hence the map

(1.21)
$$
\begin{aligned}
\partial : \quad R^{*\natural} &\rightarrow SU(2) \\
\rho &\mapsto \rho(\partial F^*)
\end{aligned}
$$

is well-defined up to conjugation by elements of $SU(2)$ (which preserves orientation). Note that $R^\natural = \partial^{-1}(1)$. Thus, near regular points of $R^\natural$, there is an identification of the normal fiber (in $R^{*\natural}$) with $\mathbf{su}(2)$, well-defined up to orientation preserving linear maps (i.e. $\mathrm{Ad}(SU(2))$). Choose $[R^{*\natural}]$ so that

(1.22)
$$
[R^{*\natural}] = [R^\natural][\mathbf{su}(2)].
$$

($[\mathbf{su}(2)]$ is, of course, the standard orientation of $\mathbf{su}(2)$.)

Now we give an orientation convention for intersections of manifolds. Let A and B be oriented properly embedded submanifolds of an oriented manifold Y, intersecting transversely in X. Let α be the normal bundle of A in Y. Orient α so that

$$
[Y] = [\alpha][A].
$$

$\alpha|_X$ is the normal bundle of X in B. Orient X so that

$$
[B] = [\alpha|_X][X].
$$

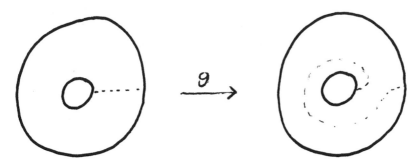

Figure 1.1: Prototypical left-handed Dehn twist.

Note that this orientation convention is compatible with the convention given above for boundaries in the sense that

$$[\partial X] = [\partial A \cap \partial B].$$

Note also that the induced orientation of X depends on the ordering of A and B.

G. Dehn Twists and Dehn Surgery.

Let $f : [1, 2] \to \mathbf{R}$ be a smooth function such that $f(r) = 0$ for r near 1 and $f(r) = 2\pi$ for r near 2. Let $A \stackrel{\text{def}}{=} \{z \in \mathbf{C} \,|\, 1 \le |z| \le 2\}$. Define $g : A \to A$ by

$$g(re^{i\theta}) \stackrel{\text{def}}{=} re^{i(\theta + f(r))}$$

for $1 \le r \le 2$ and $0 \le \theta < 2\pi$ (see Figure 1.1). Note that g is the identity near ∂A and the the isotopy class of g does not depend on the choice of f.

Let γ be a simple closed curve in an oriented surface F. Let $\varphi : A \to F$ be orientation preserving embedding of A in F which maps the boundary components of A to curves which are isotopic to γ. Define $h_\gamma : F \to F$ by

$$h_\gamma(x) = \begin{cases} \varphi g \varphi^{-1}(x), & x \in \varphi(A) \\ x, & x \notin \varphi(A). \end{cases}$$

The isotopy class of h_γ does not depend on the choice of φ. Any map in the isotopy class of h_γ is called a left-handed Dehn twist along γ. The inverses of such maps are called right-handed Dehn twists.

Let N be a 3-manifold and let $T \subset \partial N$ be a boundary component of N which is diffeomorphic to a torus. Let $a \in H_1(T; \mathbf{Z})$ be a primitive homology class (that is, one which can be represented by a simple closed curve). Let $f : \partial(S^1 \times D^2) \to T$ be a diffeomorphism which sends $\{1\} \times \partial D^2$ to a curve representing a.

$$N_a \overset{\text{def}}{=} N \cup_f (S^1 \times D^2)$$

is called the Dehn surgery of N along a. Note that N_a does not depend on the sign of a. If N is oriented, we give N_a the orientation induced from $N \subset N_a$.

If $K \subset M$ is a knot in a 3-manifold M, then Dehn surgery on K means Dehn surgery on $M \setminus U$, where U is an open tubular neighborhood of K. If $a \in H_1(\partial(M \setminus U); \mathbf{Z})$ is primitive, then $K_a \overset{\text{def}}{=} (M \setminus U)_a$.

Define a meridian of K to be a generator of

$$\ker(H_1(\partial(M \setminus U); \mathbf{Z}) \to H_1(\overline{U}; \mathbf{Z})).$$

Define a longitude of K to be a generator of

$$\ker(H_1(\partial(M \setminus U); \mathbf{Z}) \to H_1(M \setminus U; \mathbf{Z})).$$

These homology classes are well-defined up to sign. If K is a null-homologous knot (i.e. $[K] = 0 \in H_1(M; \mathbf{Z})$), then a meridian, m, and a longitude, l, of K form a basis of $H_1(\partial(M \setminus U); \mathbf{Z})$. In this case, let $K_{p/q}$ denote K_{pm+ql}, where the signs of m and l are chosen so that $\langle m, l \rangle = 1$. With this sign convention, m and l are call a standard basis of $H_1(\partial(M \setminus U); \mathbf{Z})$. (Here $\langle \cdot, \cdot \rangle$ is the intersection pairing on $H_1(\partial(M \setminus U); \mathbf{Z})$ and $\partial(M \setminus U)$ is oriented according to the "inward normal last" convention.)

§ 2

Definition of λ

In this section we define an invariant $\lambda(M)$ of an oriented rational homology sphere M. Section A contains the definition of λ. In section B, we show that $\lambda(M)$ is independent of the various choices made in section A. Notation from Section 1 will be retained throughout.

A. Definition.

Let M be a rational homology sphere. By (1.20), Q_1 and Q_2 can be put into general position via a special isotopy. We will abuse notation and continue to denote the isotoped image of Q_1 by Q_1. The same goes for η_1, θ_1, $H^1(W_1; \mathbf{h}^\perp_{\mathrm{Ad}\rho})$, etc.

$Q_1 \cap Q_2$ now consists of a finite number of points. It would be nice if we could assign a sign to each point of $Q_1 \cap Q_2$, sum these signs, and get a well-defined invariant of M. In particular, the sum would have to be special isotopy invariant, since we began with an arbitrary special isotopy. Alas, this cannot work. For consider the case where F has genus 2.

In this case the fiber of ξ is $C(S^1 \times S^1/S^1) \cong C(S^1)$ and the fiber of θ_j is $C(\mathbb{C}P^0) \cong C(\mathrm{pt})$ (see (1.8)). ($C(X)$ denotes the cone on X.) Let $p \in T_1^- \cap T_2^-$. (Such points exist if $H_1(M; \mathbf{Z})$ has elements of order greater than two.) Then the triple $(R, Q_1, Q_2) \cap \xi_p$ is diffeomorphic to $(C(S^1), C(a_1), C(a_2))$, where $a_1, a_2 \in S^1$ are distinct. There is an isotopy $h : C(S^1) \times I \to C(S^1)$, supported away from $\partial C(S^1)$, such that $h(\,\cdot\,, 0) = \mathrm{id}$, $h(\,\cdot\,, 1) = \mathrm{id}$ near the cone point, and $h(C(a_1), 1)$ intersects $C(a_2)$ once transversely (see Figure 2.1). h (or rather a suitably scaled exponential of h) extends to an isotopy h' of R supported in a neighborhood of p. Note that $h'(\,\cdot\,, 1)$ is the identity near $Q_1 \cap Q_2$. However, $h'(Q_1, 1) \cap Q_2$ has an

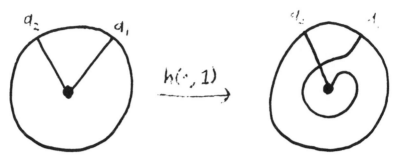

Figure 2.1: A troublesome isotopy.

extra irreducible intersection.

The above discussion makes it clear that if the irreducible points of $Q_1 \cap Q_2$ are to contribute their signs to $\lambda(M)$, the contribution of the reducible points (i.e. $T_1 \cap T_2$) must reflect global "twisting" in the normal bundle of S. In other words, we must define, for each $p \in T_1 \cap T_2$, a number $I(p)$ such that

$$\sum_{p \in Q_1^- \cap Q_2^-} \text{sign}\,(p) + \sum_{p \in T_1 \cap T_2} I(p)$$

is independent of the choice of isotopy used to put Q_1 into general position with Q_2. Defining $I(p)$ and proving isotopy invariance will occupy most of the rest of this section.

Before proceeding with the details, we give, for motivation, a rough and oversimplified sketch of the definition of $I(p)$ in the genus 2 case. As noted above, in this case $\hat{\xi}$ is an S^1 bundle over S^- and $\hat{\theta}_j$ is a section of $\hat{\xi}$ defined over T_j^-. Assume, contrary to fact, that $\hat{\xi}$ is defined over all of S and that $\hat{\theta}_j$ is defined over all of T_j. (This allows certain technicalities to be suppressed.) Let $p \in T_1^- \cap T_2^-$. Choose arcs $\alpha_j \subset T_j$ from 1 to p. Let γ be the loop $\alpha_1 * (-\alpha_2)$. The sections $\hat{\theta}_j|_{\alpha_j}$ give a section of $\hat{\xi}|_\gamma$ with discontinuities at 1 and p. These discontinuities can be repaired by moving in the positive direction (with respect to the orientation of $\hat{\xi}$) from $\hat{\theta}_1$ to $\hat{\theta}_2$ at 1 and p. This gives a trivialization Φ of $\hat{\xi}|_\gamma$. A little thought will convince one that Φ changes by ± 1 if Q_1 is modified by the troublesome isotopy of Figure 2.1. Thus if we had a canonical trivialization Φ_0 of $\hat{\xi}|_\gamma$, we could define $I(p)$ to be the difference $\Phi - \Phi_0$.

Suppose that γ bounds a surface $E \subset S$. Then $\hat{\xi}|_E$ determines a trivialization Φ_0 of $\hat{\xi}|_\gamma$, and the difference $\Phi - \Phi_0$ is just the relative first Chern class $c_1(\hat{\xi}|_E, \Phi)$. Unfortunately, if we choose a different surface E' bounded by γ, then

$$c_1(\hat{\xi}|_E, \Phi) - c_1(\hat{\xi}|_{E'}, \Phi) = c_1(\hat{\xi}|_{E \cup (-E')}),$$

which is not in general zero. But recall (Lemma (1.15)) that the symplectic form ω represents $c_1(\hat{\xi}) = c_1(\det(\nu))$. Thus

$$I(p) \overset{\text{def}}{=} c_1(\hat{\xi}|_E, \Phi) - \int_E \omega$$

is independent of the choice of spanning surface E. Using the fact that T_j is lagrangian, it is not hard to show that $I(p)$ is independent of the choice the arcs α_j.

In practice, we will carry out the above construction equivariantly in the double cover $\tilde{S}$ of S. We will also replace $\hat{\xi}$ and $\hat{\theta}_j$ with $\det^1(\nu)$ and $\det^1(\eta_j)$, since the latter bundles extend over all of $\tilde{S}$.

Now for the details.

Let $\tilde{S}$ be the double cover of S. Pick a metric on F, inducing an almost complex structure on ν (see (1.8)). Let $\det^1(\nu)$ be the unit determinant line bundle of the Zariski normal bundle ν of S^-, lifted to $\tilde{S}^-$. By (1.13), $\det^1(\nu)$ extends canonically over $\tilde{S}$. Recall from (1.13) that the Zariski normal bundle η_j of T_j^- (lifted to $\widetilde{T_j^-}$) determines a section $\det^1(\eta_j)$ of $\det^1(\nu)|_{\tilde{T}_j^-}$ which extends smoothly and canonically over all of $\widetilde{T}_j$.

Let $p \in T_1 \cap T_2$. Let $p', p'' \in \tilde{S}$ be the inverse images of p. (If $p \in P$, then $p' = p''$.) Choose arcs α_j' $[\alpha_j'']$ from 1 to p' $[p'']$ in $\widetilde{T}_j$ such that $\tau(\alpha_j') = \alpha_j''$, where τ is the covering involution of $\tilde{S}$. Let $\gamma' = \alpha_1' * (-\alpha_2')$ and $\gamma'' = \alpha_1'' * (-\alpha_2'')$.

Since $\tau_* : H_1(\tilde{S}; \mathbf{Z}) \to H_1(\tilde{S}; \mathbf{Z})$ is equal to -1 and $\tau(\gamma') = \gamma''$, γ' is homologous to $-\gamma''$. Therefore we can choose a (possibly singular) surface E in $\tilde{S}$ such that $\partial E = \gamma' \cup \gamma''$.

$\det^1(\eta_1)$ and $\det^1(\eta_2)$ are sections of $\det^1(\nu)$ over α_1' and α_2'. We wish to patch these sections together to get a trivialization of $\det^1(\nu)|_{\gamma'}$. This we do as follows. Let $x = 1$ or p'. $\eta_{1,x}$ and $\eta_{2,x}$ are transverse oriented lagrangian subspaces of ν_x (see (1.13)). Suppose we are given a path $P(\eta_{1,x}, \eta_{2,x})$ of oriented lagrangian subspaces from $\eta_{1,x}$ to $\eta_{2,x}$. Then $\det^1(P(\eta_{1,x}, \eta_{2,x}))$ is a path in $\det^1(\nu_x)$ connecting $\det^1(\eta_{1,x})$ to $\det^1(\eta_{2,x})$. These paths, together with $\det^1(\eta_j)|_{\alpha_j'}$, determine a map of S^1 into $\det^1(\nu)|_{\gamma'}$, namely

$$\det^1(\eta_1)|_{\alpha_1'} * \det^1(P(\eta_{1,p'}, \eta_{2,p'})) * (-\det^1(\eta_2)|_{\alpha_2'}) * (-\det^1(P(\eta_{1,1}, \eta_{2,1})).$$

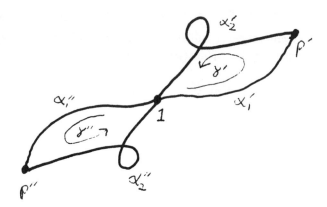

Figure 2.2: γ' and γ'' (artist's conception).

(Here the negative signs signify a reversal of path orientation.) A nonzero section of $\det^1(\nu)|_{\gamma'}$ in the same homotopy class as this map gives a trivialization of $\det^1(\nu)|_{\gamma'}$. Repeating all of the above for γ'', we get a trivialization Φ over all of ∂E.

It remains to define the path $P(\eta_{1,x}, \eta_{2,x})$. There is no unique natural choice, so instead we will define two paths, $P_+(\eta_{1,x}, \eta_{2,x})$ and $P_-(\eta_{1,x}, \eta_{2,x})$, yielding two trivializations, Φ_+ and Φ_-, and use their "average", Φ, to compute $I(p)$. (If f is an affine function from the set of trivializations to $\mathbf{R}$, then $f(\Phi) \overset{\text{def}}{=} (f(\Phi_+) + f(\Phi_-))/2$.) In what follows, we will usually pretend that Φ is an honest trivialization.

Let L_0, L_1 be transverse, oriented lagrangians in a symplectic vector space. Let $e_1, \ldots, e_n$ be a oriented basis of L_0. Then there is a unique basis $f_1, \ldots, f_n$ of L_1 such that $\omega(e_i, f_j) = \delta_{ij}$. (Note that if $L_{j-1} = \eta_{j,x}$ with the J-complex orientation, then this basis is oriented correctly.) Let $P_\pm(L_0, L_1)_t$ be the (oriented) span of $g_{1,t}, \ldots, g_{n,t}$, where

$$g_{i,t} = (1-t)e_i \pm (t-1)f_i.$$

(Note that $P_\pm(L_0, L_1)_t$ is transverse to L_0 and L_1 for all $0 < t < 1$. This fact will be used later.) Since the path $P_\pm(L_0, L_1)$ depends continuously on the choice of oriented basis for L_0, and the set of all such bases is connected, we see that the homotopy class of $P_\pm(L_0, L_1)$ is well defined. Note that changing the sign of ω interchanges P_+ and P_-, and so preserves Φ.

Now define

$$(2.1) \qquad I(p) \overset{\text{def}}{=} \begin{cases} \frac{1}{2}(c_1(\det^1(\nu)|_E, \Phi) - \int_E \omega) & p \notin P \\ \frac{1}{4}(c_1(\det^1(\nu)|_E, \Phi) - \int_E \omega) & p \in P. \end{cases}$$

It must be shown that $I(p)$ does not depend on the choice of spanning surface, arcs, or the metric on F.

Suppose we choose a different surface E' such that $\partial E' = \gamma' \cup \gamma''$. Then $I(p)$ changes (up to a scalar) by

$$c_1(\det^1(\nu)|_{E'}, \Phi) - \int_{E'} \omega - c_1(\det^1(\nu)|_E, \Phi) + \int_E \omega$$

$$= c_1(\det^1(\nu)|_{E' \cup (-E)}) - \int_{E' \cup (-E)} \omega$$

$$= 0.$$

(Here we have used (1.15).)

Suppose we replace α_1' with $\hat{\alpha}_1'$, another arc from 1 to p' in $\widetilde{T_1}$. (We of course also replace $\alpha_1'' = \tau(\alpha_1')$ with $\hat{\alpha}_1'' = \tau(\hat{\alpha_1'})$.) Let $\hat{\gamma}' = \hat{\alpha}_1' * (-\alpha_2')$ and $\hat{\gamma}'' = \hat{\alpha}_1'' * (-\alpha_2'')$. Let δ' be the loop $\hat{\alpha}_1' * (-\alpha_1')$ and let $\delta'' = \tau\delta' = \hat{\alpha}_1'' * (-\alpha_1'')$. Let H be a surface in $\widetilde{T_1}$ such that $\partial H = \delta' \cup \delta''$. Then $\partial(E \cup H) = \hat{\gamma}' \cup \hat{\gamma}''$, and we can use $E \cup H$ in place of E to compute the new $I(p)$. Let $\hat{\Phi}$ be the trivialization of $\det^1(\nu)|_{\partial(E \cup H)}$ determined by $\det^1(\eta_j)$.

Since $H \subset \widetilde{T_1}$ and $\widetilde{T_1}$ is lagrangian with respect to ω,

$$\int_{E \cup H} \omega = \int_E \omega.$$

Since $\det^1(\eta_1)$ is defined over all of H,

$$c_1(\det^1(\nu)|_{E \cup H}, \hat{\Phi}) = c_1(\det^1(\nu)|_E, \Phi).$$

Thus $I(p)$ is independent of the choice of α_1' and α_1''. Similarly, is independent of the choice of α_2' and α_2''.

The choice of metric on F clearly does not affect the $\int_E \omega$ part of $I(p)$. The c_1 part depends continuously on the metric and takes discrete values. Since the space of metrics on F is connected, this part is also independent of the choice of metric.

Now define

$$(2.2) \qquad \langle Q_1, Q_2 \rangle \overset{\text{def}}{=} \sum_{p \in Q_1^- \cap Q_2^-} \text{sign}(p) + \sum_{p \in T_1 \cap T_2} I(p).$$

It will be useful to have a slightly different viewpoint of the above at our disposal. Define a *wiring* of Q_1 and Q_2 to be a collection of (homotopy classes of) arcs α'_j, α''_j and surfaces E, as above, for each point of $T_1 \cap T_2$. Given a wiring $\mathcal{W}$, define (for each $p \in T_1 \cap T_2$)

$$(2.3) \qquad I_1(p,\mathcal{W}) \stackrel{\text{def}}{=} \begin{cases} \frac{1}{2}(c_1(\det^1(\nu)|_E, \Phi)) & p \notin P \\ \frac{1}{4}(c_1(\det^1(\nu)|_E, \Phi)) & p \in P \end{cases}$$

and

$$I_2(p,\mathcal{W}) \stackrel{\text{def}}{=} \begin{cases} \frac{1}{2}\left(-\int_E \omega\right) & p \notin P \\ \frac{1}{4}\left(-\int_E \omega\right) & p \in P. \end{cases}$$

Hence

$$I(p) = I_1(p,\mathcal{W}) + I_2(p,\mathcal{W}).$$

Define

$$(2.4) \quad A_1(Q_1,Q_2,\mathcal{W}) \stackrel{\text{def}}{=} \sum_{p \in Q_1^- \cap Q_2^-} \text{sign}\,(p) + \sum_{p \in T_1 \cap T_2} I_1(p,\mathcal{W})$$

$$A_2(Q_1,Q_2,\mathcal{W}) \stackrel{\text{def}}{=} \sum_{p \in T_1 \cap T_2} I_2(p,\mathcal{W})$$

Thus we have

$$(2.5) \qquad \langle Q_1, Q_2 \rangle = A_1(Q_1,Q_2,\mathcal{W}) + A_2(Q_1,Q_2,\mathcal{W}).$$

Finally, define

$$(2.6) \qquad \lambda(M) \stackrel{\text{def}}{=} \frac{\langle Q_1, Q_2 \rangle}{|H_1(M;\mathbf{Z})|}.$$

Note that since $I(1) = 0$, this definition of λ agrees with Casson's in the case where M is a $\mathbf{Z}$HS, up to a factor of $1/2$. (That the signs agree is not clear at this point, but will become so later (see (3.50)). In particular, $\lambda(M) = 0$ if $\pi_1(M) = 1$.

B. Well-definition.

We now show that the right hand side of (2.6) does not depend on the choice of general positioning isotopy, the choice of orientations of $\widetilde{T_1}$ and $\widetilde{T_2}$, or the choice of Heegaard splitting of M. This will show that $\lambda(M)$ is well-defined.

First we show that $\lambda(M)$ is independent of the choice of special isotopy used to put Q_1 and Q_2 into general position. This is equivalent to showing that, given a special isotopy $h : R \times I \rightarrow R$ ($h(\,\cdot\,,0) = \mathrm{id}$) such that $Q_1^\dagger \overset{\text{def}}{=} h(Q_1, 1)$ is in general position with Q_2, we have

$$(2.7) \quad \sum_{p \in Q_1^- \cap Q_2^-} \operatorname{sign}(p) + \sum_{p \in T_1 \cap T_2} I(p) = \sum_{p \in Q_1^{\dagger -} \cap Q_2^-} \operatorname{sign}(p) + \sum_{p \in T_1 \cap T_2} I^\dagger(p),$$

where $I^\dagger(p)$ is computed just as $I(p)$, but with Q_1, η_1, etc. replaced by $Q_1^\dagger$, $\eta_1^\dagger$, etc. (Note that we are continuing to assume that Q_1 and Q_2 have already been isotoped into general position.)

It will be useful to introduce some auxiliary spaces. Let $\hat{\xi} \subset \xi$ be the unit normal bundle of S^- in R. Consider the map

$$e : \quad \hat{\xi} \times I \;\rightarrow\; R$$
$$(x, t) \;\longmapsto\; \exp(tx).$$

(Here exp denotes some diffeomorphism from a neighborhood of the zero section of ξ to a neighborhood of S^- in R whose differential at the zero section is the identity.) There is a neighborhood N of $\hat{\xi} \times \{0\}$ such that e restricted to $N \backslash \hat{\xi} \times \{0\}$ is a diffeomorphism onto its image. Define

$$\overline{R^-} \overset{\text{def}}{=} N \cup_e R^-.$$

Define $\overline{Q_j^-} \subset \overline{R^-}$ similarly. Note that $\overline{R^-}\,[\overline{Q_j^-}]$ is a manifold with boundary $\hat{\xi} \times \{0\}\,[\hat{\theta}_j \times \{0\}]$. h induces a map (also denoted h) $h : \overline{R^-} \times I \rightarrow \overline{R^-}$. We will identify $\overline{Q_1^-} \times I$ with its image under h.

By a general position argument, we may assume that $\overline{Q_1^-} \times I$ and $\overline{Q_2^-}$ are transverse. It follows that $(\overline{Q_1^-} \times I) \cap \overline{Q_2^-}$ is a properly embedded 1-manifold, and that $C \overset{\text{def}}{=} \partial(\overline{Q_1^-} \times I \cap \overline{Q_2^-})$ consists of $Q_1^- \cap Q_2^-$, $Q_1^{\dagger -} \cap Q_2^-$ and $(\hat{\theta}_1 \times I) \cap \hat{\theta}_2$. (Here we have used the fact that Q_1 is transverse to Q_2 at P.) Orient $\overline{Q_1^-} \times I$ as a product. Orient $(\overline{Q_1^-} \times I) \cap \overline{Q_2^-}$ as an intersection (see (1.F)). For $c \in C$, define $s(c)$ to be $+1$ $[-1]$ if the inward pointing normal at c is positively [negatively] oriented.

With the above orientation conventions, we have

$$\sum_{p \in Q_1^- \cap Q_2^-} \operatorname{sign}(p) - \sum_{p \in Q_1^{\dagger -} \cap Q_2^-} \operatorname{sign}(p) \;=\; \sum_{c \in Q_1^- \cap Q_2^-} s(c) + \sum_{c \in Q_1^{\dagger -} \cap Q_2^-} s(c)$$

$$(2.8) \qquad\qquad\qquad\qquad\qquad\qquad =\; -\sum_{c \in (\hat{\theta}_1 \times I) \cap \hat{\theta}_2} s(c).$$

In view of (2.8), (2.7) reduces to

$$(2.9) \qquad \sum_{p \in T_1 \cap T_2} (I(p) - I^\dagger(p)) = \sum_{c \in (\hat{\theta}_1 \times I) \cap \hat{\theta}_2} s(c).$$

Since special isotopies are fiber preserving and keep Q_1 and Q_2 Zariski transverse at P, we have

$$(\hat{\theta}_1 \times I) \cap \hat{\theta}_2 = \bigcup_{p \in T_1^- \cap T_2^-} (\hat{\theta}_{1,p} \times I) \cap \hat{\theta}_{2,p}.$$

Furthermore, it is easy to see that if $p \in T_1 \cap T_2 \cap P$, then $I(p) = I^\dagger(p)$. Thus, in order to prove (2.9) it suffices to show that

$$(2.10) \qquad I(p) - I^\dagger(p) = \sum_{c \in (\hat{\theta}_{1,p} \times I) \cap \hat{\theta}_{2,p}} s(c)$$

for all $p \in T_1^- \cap T_2^-$.

First we analyze the right hand side of (2.10). Let $c \in (\hat{\theta}_{1,p} \times I) \cap \hat{\theta}_{2,p}$. Let $p' \in \widetilde{T}_1 \cap \widetilde{T}_2$ be an inverse image of p. Keeping in mind the orientation conventions of (1.F), we see that

$$
\begin{aligned}
s(c) &= \frac{[\partial(\overline{Q_1^-} \times I)][\partial \overline{Q_2^-}]}{[\partial \overline{R^-}]} \\[2mm]
&= -\frac{[\partial \overline{Q_1^-}][I][\partial \overline{Q_2^-}]}{[\partial \overline{R^-}]} \\[2mm]
&= -\frac{[\widetilde{T}_1][\hat{\theta}_1][I][\widetilde{T}_2][\hat{\theta}_2]}{[\widetilde{S}][\hat{\xi}]} \\[2mm]
&= (-1)^{g-1} \frac{[\hat{\theta}_1][I][\hat{\theta}_2]}{[\hat{\xi}]} \frac{[\widetilde{T}_1][\widetilde{T}_2]}{[\widetilde{S}]} \\[2mm]
&= (-1)^{g-1} \frac{[\hat{\theta}_1][I][\hat{\theta}_2]}{[\hat{\xi}]}.
\end{aligned}
$$

(All juxtapositions of orientations are considered to take place at c or p'.) Therefore

$$(2.11) \qquad \sum_{c \in (\hat{\theta}_{1,p} \times I) \cap \hat{\theta}_{2,p}} s(c) = (-1)^{g-1} \langle (\hat{\theta}_{1,p'} \times I), \hat{\theta}_{2,p'} \rangle_{\hat{\xi}_{p'}}.$$

Let $\beta : I \to \mathcal{L}_{p'}^C$ be a path of complex lagrangians, transverse to $\eta_{2,p'}$, from $\eta_{1,p'}^\dagger$ to $\eta_{1,p'}$. (This is possible, since that space of complex lagrangians transverse to a fixed one is contractable.) Let δ be the loop

$$(2.12) \qquad\qquad (\eta_{1,p'} \times I) * \beta.$$

Since $(\hat{\theta}_{1,p'} \times I) = CP(\eta_{1,p'} \times I)$ and $CP(\beta)$ is disjoint from $\hat{\theta}_{2,p'} = CP(\eta_{2,p'})$, (2.11) becomes

$$\sum_{c \in (\hat{\theta}_{1,p} \times I) \cap \hat{\theta}_{2,p}} s(c) \;=\; (-1)^{g-1} \langle CP(\delta), CP(\eta_{2,p'}) \rangle_{\hat{\xi}_{p'}}$$

$$(2.13) \qquad\qquad\qquad\qquad = (-1)^{g-1} \psi([\delta]).$$

(Recall that $\psi : \pi_1(\mathcal{L}_{p'}^C) \to \mathbf{Z}$ was defined in (1.10).)

Now for the left hand side of (2.10). First of all, note that since special isotopies are fixed on S, we can use the same wiring $\mathcal{W}$ for computing $I(p)$ and $I^\dagger(p)$. It follows that

$$(2.14) \qquad I(p) - I^\dagger(p) = I_1(p, \mathcal{W}) - I_1^\dagger(p, \mathcal{W}) = -\frac{1}{2}(\Phi - \Phi^\dagger).$$

(Φ is computed using $\eta_1 \times \{0\} = \eta_1$, $\Phi^\dagger$ is computed using $\eta_1 \times \{1\} = \eta_1^\dagger$.) Note that

$$\Phi - \Phi^\dagger = (\Phi|_{\gamma'} - \Phi^\dagger|_{\gamma'}) + (\Phi|_{\gamma''} - \Phi^\dagger|_{\gamma''}).$$

Let $\Psi : \det^1(\nu)|_{\gamma'} \to S^1$ be a trivialization. Then

$$\Phi|_{\gamma'} - \Phi^\dagger|_{\gamma'} = \deg(\Psi(a)) - \deg(\Psi(a^\dagger)),$$

where

$$a = (-\det^1(\eta_2)|_{\alpha_2'}) * (-\det^1(P(\eta_{1,1}, \eta_{2,1}))) *$$
$$(\det^1(\eta_1)|_{\alpha_1'}) * (\det^1(P(\eta_{1,p'}, \eta_{2,p'})))$$
$$a^\dagger = (-\det^1(\eta_2)|_{\alpha_2'}) * (-\det^1(P(\eta_{1,1}^\dagger, \eta_{2,1}))) *$$
$$(\det^1(\eta_1^\dagger)|_{\alpha_1'}) * (\det^1(P(\eta_{1,p'}^\dagger, \eta_{2,p'}))).$$

(Here deg denotes the degree of a map from S^1 to S^1, and the negative signs indicate that a path should be traversed in the reverse direction.) Consider the 1-parameter family of paths

$$(-\det^1(\eta_2)|_{\alpha_2'}) * (-\det^1(P(\eta_{1,1} \times \{t\}, \eta_{2,1}))) * (\det^1(\eta_1 \times \{t\})|_{\alpha_1'})$$

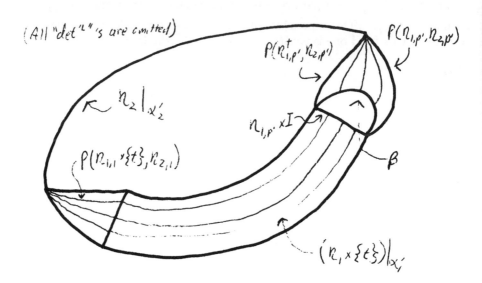

Figure 2.3: Various paths in $\det^1(\nu)|_{\gamma'}$.

$(0 \leq t \leq 1)$. Adding the boundary of the above square to $a \cup a^{\dagger}$, we see that

$$\Phi|_{\gamma'} - \Phi^{\dagger}|_{\gamma'} = \deg(\Psi(b)),$$

where

$$b = (-\det^1(\eta_{1,p'} \times I)) * (\det^1(P(\eta_{1,p'}, \eta_{2,p'}))) * (-\det^1(P(\eta_{1,p'}^{\dagger}, \eta_{2,p'}^{\dagger}))).$$

(Here $\det^1(\eta_{1,p'} \times I)$ is viewed as a path from $\det^1(\eta_{1,p'})$ to $\det^1(\eta_{1,p'}^{\dagger})$ in $\det^1(\nu_{p'})$.)

Recall the path β of complex lagrangians transverse to $\eta_{2,p'}$ from $\eta_{1,p'}^{\dagger}$ to $\eta_{1,p'}$. The 1-parameter family of paths $P(\beta_t, \eta_{2,p'})$ can be viewed as a homotopy (rel end points) from

$$(\det^1(P(\eta_{1,p'}, \eta_{2,p'}))) * (-\det^1(P(\eta_{1,p'}^{\dagger}, \eta_{2,p'})))$$

to $-\det^1(\beta)$. Hence

$$\Phi|_{\gamma'} - \Phi^{\dagger}|_{\gamma'} = \deg(\det^1(-\delta)) = -\varphi([\delta]),$$

where δ is as defined in (2.12) and φ is the homomorphism defined in (1.10). Note that Ψ can be (and has been) omitted, since everything is taking place in the single fiber $\det^1(\nu_{p'})$.

By equivariance, we also have

$$\Phi|_{\gamma''} - \Phi^\dagger|_{\gamma''} = -\varphi([\delta]),$$

and hence
(2.15) $$I(p) - I^\dagger(p) = \varphi([\delta]).$$

(2.10) now follows from (2.13), (2.15) and (1.12), and the proof of isotopy invariance is complete.

In the above argument, the fact that the isotopy was fixed on S was needed only to show that the A_2 part of $\langle Q_1, Q_2 \rangle$ was invariant. The A_1 part of $\langle Q_1, Q_2 \rangle$ is invariant under a less restricted class of isotopies, and this fact will be used in the proof of (3.36).

More precisely, define a *not-so-special isotopy* $\{h_t\}_{0 \le t \le 1}$ of R to be one which satisfies

1. $h_t(Q_1)$ is transverse to Q_2 at $Q_1 \cap Q_2 \cap P$ for all t.

2. $h_t(T_1)$ is transverse to T_2 in S for all t.

3. h_t induces a symplectic bundle map of ν for all t.

Let $\mathcal{W}$ be a wiring of Q_1 and Q_2 and $\{h_t\}$ be a not-so-special isotopy. Then condition 2 above allows us to choose wirings $h_t(\mathcal{W})$ of $h_t(Q_1)$ and Q_2 so that the arcs and surfaces of $h_t(\mathcal{W})$ vary continuously with t. An argument similar to the one used to establish the special isotopy invariance of $\langle Q_1, Q_2 \rangle$ shows

(2.16) *Let* $\mathcal{W}$ *be a wiring of* Q_1 *and* Q_2 *and* $\{h_t\}$ *be a not-so-special isotopy. Then* $A_1(h_1(Q_1), Q_2, h_1(\mathcal{W})) = A_1(Q_1, Q_2, \mathcal{W})$.

Suppose that the orientations of $\widetilde{T_1}$ and $\widetilde{T_2}$ are both changed. Then, according to the orientation conventions of (1.F), the orientations of Q_1 and Q_2 also change, while all other orientations remain the same. It is now easy to see that these changes do not affect (2.6).

Finally, it must be shown that (2.6) does not depend on the choice of Heegaard splitting for M. First we recall the notion of a stabilization of a Heegaard splitting. Let (V_1, V_2, Σ) be the standard genus one Heegaard splitting of S^3. Let

$$(W_1^\natural, W_2^\natural, F^\natural) \overset{\text{def}}{=} (W_1, W_2, F) \# (V_1, V_2, \Sigma).$$

$(W_1^\natural, W_2^\natural, F^\natural)$ is called a stabilization of (W_1, W_2, F). We can make the identifications

$$
\begin{aligned}
F^\natural &= F \# \Sigma \\
W_j^\natural &= W_j \#_\partial V_j
\end{aligned}
$$

(where $\#_\partial$ denotes the boundary connected sum). See [AM] for more details.

By [S] (see also [AM]), any two Heegaard splittings of M are equivalent after a sufficient number of stabilizations. Thus it suffices to show that (2.6) does not change when (W_1, W_2, F) is stabilized to $(W_1^\natural, W_2^\natural, F^\natural)$.

Use the superscript $\natural$ to denote objects associated to the stabilized Heegaard splitting. (e.g. $R^\natural = \mathrm{Hom}(\pi_1(F^\natural), SU(2))/SU(2)$.) Let F^- be F minus a disk. Define Σ^- similarly. Make the identifications

$$
\begin{aligned}
F^\natural &= F^- \cup \Sigma^- \\
F &= F^- \cup D^2 \\
\Sigma &= \Sigma^- \cup D^2.
\end{aligned}
$$

$\pi_1(F^\natural)$ is generated by $\pi_1(F^-)$ and $\pi_1(\Sigma^-)$. We can identify

$$
R = \{[\rho] \in R^\natural \mid \rho(\pi_1(\Sigma^-)) = \{1\}\}.
$$

Similarly,

$$
Q_j = \{[\rho] \in Q_j^\natural \mid \rho(\pi_1(V_j)) = \{1\}\}.
$$

It is easy to see that

$$
(2.17) \qquad Q_1^\natural \cap Q_2^\natural = Q_1 \cap Q_2.
$$

Our first task is to show that the special isotopies of R used to put Q_1 and Q_2 into general position can be extended to $R^\natural$ so that (2.17) remains true. Toward this end, we analyze the normal bundles of R in $R^\natural$ and Q_j in $Q_j^\natural$.

Let $\gamma \overset{\text{def}}{=} \partial F^- = \partial \Sigma^-$. Let $[\rho] \in R$. By examining the Mayer-Vietoris sequence for the triple (F, F^-, D^2) with coefficients in the flat vector bundle $\mathbf{g}_{\mathrm{Ad}\rho}$, one finds that

$$
T_{[\rho]} R = H^1(F; \mathbf{g}_{\mathrm{Ad}\rho}) = \ker(H^1(F^-; \mathbf{g}_{\mathrm{Ad}\rho}) \to H^1(\gamma; \mathbf{g}_{\mathrm{Ad}\rho})).
$$

The Mayer-Vietoris sequence for $(F^\natural, F^-, \Sigma^-)$ yields

$$
0 \to H^1(F^\natural; \mathbf{g}_{\mathrm{Ad}\rho}) \to H^1(F^-; \mathbf{g}_{\mathrm{Ad}\rho}) \oplus H^1(\Sigma^-; \mathbf{g}_{\mathrm{Ad}\rho}) \to H^1(\gamma; \mathbf{g}_{\mathrm{Ad}\rho}).
$$

Since $\mathbf{g}_{\mathrm{Ad}\rho}$ is trivial over Σ^-, the map $H^1(\Sigma^-;\mathbf{g}_{\mathrm{Ad}\rho}) \to H^1(\gamma;\mathbf{g}_{\mathrm{Ad}\rho})$ is zero and

$$
\begin{aligned}
T_{[\rho]}R^{\natural} &= H^1(F^{\natural};\mathbf{g}_{\mathrm{Ad}\rho}) \\
&= H^1(F;\mathbf{g}_{\mathrm{Ad}\rho}) \oplus H^1(\Sigma^-;\mathbf{g}_{\mathrm{Ad}\rho}) \\
&= T_{[\rho]}R \oplus H^1(\Sigma^-;\mathbf{g}).
\end{aligned}
$$

Therefore the Zariski normal bundle of R in $R^{\natural}$ can be identified with the trivial bundle $R \times H^1(\Sigma^-;\mathbf{g})$. It is not hard to see that the splitting $T_{[\rho]}R \oplus H^1(\Sigma^-;\mathbf{g})$ is an orthogonal direct sum of symplectic vector spaces. Similarly, the Zariski normal bundle of Q_j in $Q_j^{\natural}$ can be identified with the trivial bundle $Q_j \times H^1(V_j;\mathbf{g})$. Since $H^1(V_1;\mathbf{g})$ and $H^1(V_2;\mathbf{g})$ are transverse inside $H^1(\Sigma^-;\mathbf{g})$, the normal fibers of Q_1 and Q_2 are transverse inside the normal fiber of R at all points of $Q_1 \cap Q_2$.

It is now easy to see that a special isotopy of Q_1 in R can be extended to a special isotopy of $Q_1^{\natural}$ in $R^{\natural}$ so that (2.17) and the product structure of the normal bundle of R in $R^{\natural}$ are preserved.

Assume now that $Q_1^{\natural}$ and $Q_2^{\natural}$ have been put into general position as above. If it can be shown that for each $p \in T_1 \cap T_2$, $|I^{\natural}(p)| = |I(p)|$, and that for each $p \in Q_1^- \cap Q_2^-$,

$$
(2.18) \qquad \qquad \mathrm{sign}\,^{\natural}(p) = \mathrm{sign}\,(p),
$$

then, using the isotopy invariance of $\langle Q_1^{\natural}, Q_2^{\natural} \rangle$, it will follow that

$$
\langle Q_1^{\natural}, Q_2^{\natural} \rangle = \langle Q_1, Q_2 \rangle.
$$

First we show that $|I^{\natural}(p)| = |I(p)|$ for $p \in T_1 \cap T_2 = T_1^{\natural} \cap T_2^{\natural}$. We can use the same arcs α_j', α_j'' and surface E to compute $I^{\natural}(p)$ and $I(p)$. Since $\omega^{\natural}|_{\widetilde{S}} = \omega$, $\int_E \omega^{\natural} = \int_E \omega$. Let $\bar{\nu}$ be the bundle $\widetilde{S} \times H^1(\Sigma^-;\mathbf{h}^{\perp}_{\mathrm{Ad}\rho})$. Then $\nu^{\natural}|_{\widetilde{S}} = \nu \oplus \bar{\nu}$. Hence $\det(\nu^{\natural})|_{\widetilde{S}} = \det(\nu) \otimes \det(\bar{\nu})$. Similarly, $\det(\eta_j^{\natural})|_{\widetilde{T_j}} = \det(\eta_j) \otimes \det(\bar{\eta}_j)$, where $\bar{\eta}_j \stackrel{\mathrm{def}}{=} \widetilde{T_j} \times H^1(W_j;\mathbf{h}^{\perp})$.

Let Y_1 and Y_2 be symplectic vector spaces. Let $K_i, L_i \subset Y_i$ be oriented transverse lagrangian subspaces. It is easy to see from the definition of the path $P_{\pm}$ that

$$
P_{\pm}(K_1 \oplus K_2, L_1 \oplus L_2) = P_{\pm}(K_1, L_1) \oplus P_{\pm}(K_2, L_2).
$$

Using the canonical trivialization of $\det(\bar{\nu})$, we can identify $\det^1(\nu^{\natural})$ with $\det^1(\nu)$. Using $P_{\pm}$ to patch together the sections $\det^1(\bar{\eta}_j)$ of $\det^1(\bar{\nu})$, one gets a trivialization $\bar{\Phi}_{\pm}$ of $\det^1(\bar{\nu})|_{\partial E}$. It follows from the above remarks that $\Phi_\pm^\natural = \Phi_\pm \otimes \bar{\Phi}_\pm$. So it suffices to show that $\bar{\Phi}_\pm$ is in the same homotopy class as the canonical trivialization of $\det^1(\bar{\nu})|_{\partial E}$. This is obvious, since $\det^1(\bar{\eta}_j)$ is constant with respect to this trivialization.

Next we establish (2.18). Let $p \in Q_1^- \cap Q_2^-$. Let $K_j \overset{\text{def}}{=} H^1(V_j; \mathbf{g})$, $L \overset{\text{def}}{=} H^1(\Sigma^-; \mathbf{g})$. Orient L as a symplectic vector space and choose orientations of K_1 and K_2 so that

$$[L] = (-1)^{g-1}[K_1][K_2].$$

It follows from the orientation conventions of (1.F) (though not without some effort) that

$$[Q_j^\natural] = [Q_j][K_j].$$

Hence

$$
\begin{aligned}
\operatorname{sign}^\natural(p) &= \frac{[Q_1^\natural][Q_2^\natural]}{[R^\natural]} \\
&= \frac{[Q_1][K_1][Q_2][K_2]}{[R][L]} \\
&= (-1)^{3(3g-3)} \frac{[Q_1][Q_2]}{[R]} \frac{[K_1][K_2]}{[L]} \\
&= \frac{[Q_1][Q_2]}{[R]} \\
&= \operatorname{sign}(p).
\end{aligned}
$$

This completes the proof that λ is well-defined.

§ 3

Various Properties of λ

In this section we prove various technical lemmas needed for the proof of the general Dehn surgery formula of the next section. Many of the proofs are easy generalizations of Casson's work in the $\mathbf{Z}$HS case, and this has been indicated by the words "Casson for $\mathbf{Z}$HS case".

Throughout this section, retain all notation from the previous two sections.

(3.1) **Lemma.** *Let M be a $\mathbf{Q}$HS. Let $-M$ denote M with the opposite orientation. Then*

$$\lambda(-M) = \lambda(M).$$

Proof: (Casson for $\mathbf{Z}$HS case.) The only parts of the constructions used for defining $\lambda(M)$ which are sensitive to the orientation of M are the orientations of various spaces and the sign of ω. (In particular, the "patching procedure" used over points of $T_1 \cap T_2$ does not depend on the sign of ω.) Hence, for $p \in T_1 \cap T_2$, the absolute value of $I(p)$ does not depend on the orientation of M. It now follows from (2.2) and the isotopy invariance of $\langle Q_1, Q_2 \rangle$ that absolute value of $\langle Q_1, Q_2 \rangle$ is independent of the orientation of M.

All that is left to do is check signs. It suffices to do this at a single irreducible point of $Q_1 \cap Q_2$. Changing the orientation of M changes the orientation of F which changes the sign of ω. Changing the sign of ω changes $[S]$ by $(-1)^g$ and $[R]$ by $(-1)^{3g-3}$. This means that $[\widetilde{T_1}]$ (say), and hence $[Q_1]$, change by $(-1)^g$. Thus $\text{sign}(p)$ changes by $(-1)^{g+3g-3} = -1$.

41

□

The following lemma is the prototype for most of the others in this section.

(3.2) Lemma. *Let* γ *be a separating simple closed curve on* F. *Let* $h :$ $F \to F$ *be a left-handed Dehn twist along* γ. *Let* $M_n = W_1 \cup_{h^n} W_2$. *(That is, the gluing map* $\partial W_1 \to \partial W_2$ *is composed with* h^n.*)* *Then*

$$\lambda(M_n) = \lambda(M) + n\lambda'(\gamma)$$

for some $\lambda'(\gamma) \in \mathbf{Q}$, *independent of* n.

Proof: (Casson for $\mathbf{Z}$HS case.) h induces maps of $\pi_1(F)$, R and $R^\sharp$, which will also be denoted by h. It is clear that

$$\lambda(M_n) = \frac{\langle h^n(Q_1), Q_2 \rangle}{|H_1(M;\mathbf{Z})|}.$$

(Here we have used the fact that $H_1(M_n;\mathbf{Z}) \cong H_1(M;\mathbf{Z})$ for all n.) So what we must show is that

$$\lambda(M_{n+1}) - \lambda(M_n) = \frac{(\langle h^{n+1}(Q_1), Q_2 \rangle - \langle h^n(Q_1), Q_2 \rangle)}{|H_1(M;\mathbf{Z})|} = \lambda'(\gamma)$$

for some $\lambda'(\gamma) \in \mathbf{Q}$, independent of n.

The idea of the proof is to represent the difference $h(Q_1) - Q_1$ as a cycle $D \subset R^-$. It will follow that

$$(3.3) \qquad \langle h^{n+1}(Q_1), Q_2 \rangle - \langle h^n(Q_1), Q_2 \rangle = \langle h^n(D), Q_2 \rangle,$$

where the right hand side denotes a homological intersection number. The proof will be complete upon showing that $h(D)$ is homologous to D.

Our first task is to construct the difference cycle D. Let F_1 and F_2 be the components of F cut along γ. Then $\pi_1(F)$ is the free product of $\pi_1(F_1)$ and $\pi_1(F_2)$ amalgamated over $\pi_1(\gamma)$. If the location of the base point is chosen properly, then h is given by

$$h(\alpha) = \begin{cases} \alpha, & \alpha \in \pi_1(F_1) \\ \gamma\alpha\gamma^{-1}, & \alpha \in \pi_1(F_2). \end{cases}$$

Correspondingly, for $\rho \in R^\sharp$ we have

$$h(\rho)\alpha = \begin{cases} \rho(\alpha), & \alpha \in \pi_1(F_1) \\ \rho(\gamma)\rho(\alpha)\rho(\gamma)^{-1}, & \alpha \in \pi_1(F_2). \end{cases}$$

This map descends to one on R.

There is a open ball $U \subset \mathbf{su}(2)$ such that $\exp : U \rightarrow SU(2) \backslash \{-1\}$ is a diffeomorphism. For $A = \exp(a) \in \exp(U)$ and $t \in [0,1]$, define

$$(3.4) \qquad\qquad A^t \overset{\text{def}}{=} \exp(ta).$$

Also define $(-1)^0 \overset{\text{def}}{=} 1$. Let $V_\epsilon \subset SU(2)$ be the closed ball of radius ϵ (with respect to some Ad-invariant metric) centered at $-1 \in SU(2)$. Let

$$\begin{aligned} N^\natural &= \{\rho \in R^\natural \,|\, \rho(\gamma) = -1\} \\ N_\epsilon^\natural &= \{\rho \in R^\natural \,|\, \rho(\gamma) \in V_\epsilon\}. \end{aligned}$$

Let $f : SU(2) \rightarrow [0,1]$ be a smooth, Ad-invariant, monotonic (on $SU(2)/\text{Ad}$) function such that $f = 1$ outside V_ϵ and $f(-1) = 0$. For $t \in [0,1]$, define $h_t : R^\natural \rightarrow R^\natural$ by

$$h_t(\rho)\alpha = \begin{cases} \rho(\alpha), & \alpha \in \pi_1(F_1) \\ \rho(\gamma)^{f(\rho(\gamma))t}\rho(\alpha)\rho(\gamma)^{-f(\rho(\gamma))t}, & \alpha \in \pi_1(F_2). \end{cases}$$

$\{h_t\}$ descends to an isotopy of R, also denoted $\{h_t\}$. Note that $h_0 = \text{id}$ and $h_1 = h$ on $R \backslash N_\epsilon$.

We now show that $\{h_t\}$ is a special isotopy (see (1.19)). Since $[\gamma]$ lies in the commutator subgroup of $\pi_1(F)$, the differential of the map

$$\begin{aligned} R^\natural &\rightarrow SU(2) \\ \rho &\mapsto \rho(\gamma) \end{aligned}$$

is 0 at $P^\natural$. It follows that $h_{t*} : T_p R \rightarrow T_p R$ is the identity for $p \in P$. Since $\rho(\gamma) = 1$ for $\rho \in S^\natural$, $h_t|_S = \text{id}$. It follows from Theorem 4.5 of [G3] that h_t is a symplectic map on all of R, a fortiori on $TR|_{S^-}$. Since $h_{t*} : TS^- \rightarrow TS^-$ is the identity, h_{t*} must preserve the fibers of ν.

By the special isotopy invariance proved in (2.C),

$$\langle Q_1, Q_2 \rangle = \langle h_1(Q_1), Q_2 \rangle.$$

We may assume without loss of generality that Q_1 in transverse to ∂N_ϵ. Let D be the cycle

$$(-h_1(Q_1 \cap N_\epsilon)) \cup h(Q_1 \cap N_\epsilon).$$

Note that $h_1(Q_1)$ coincides with $h(Q_1)$ on $R \backslash N_\epsilon$. Thus it follows from (2.2) that

$$(3.5) \qquad\qquad \langle h(Q_1), Q_2 \rangle - \langle Q_1, Q_2 \rangle = \langle D, Q_2 \rangle,$$

where the right hand side is to be regarded as the homological intersection number of $[D] \in H_{3g-3}(R^-; \mathbf{Z})$ and $[Q_2] \in H_{3g-3}(R, S; \mathbf{Z})$. (3.3) clearly follows from (3.5).

Now we must show that $h_*([D]) = [D] \in H_{3g-3}(R^-; \mathbf{Z})$. This follows from the facts that $D \subset N_\epsilon$, $h|_N = \mathrm{id}$, and N_ϵ deformation retracts onto N. It will be useful in what follows, however, to have an explicit representation of $[D]$ as a cycle in N. This construction occupies the remainder of the proof.

Let
$$R_i^* = \mathrm{Hom}(\pi_1(F_i), SU(2))/SU(2)$$
$(i = 1, 2)$. There is a natural map
$$\pi : R \to R_1^* \times R_2^*$$
which is a fibering over the image of N. I claim that

(3.6) $$[D] = 2[\pi^{-1}(\pi(Q_1 \cap N))] \in H_{3g-3}(R^-; \mathbf{Z}).$$

(Note that the sign of the right hand side is ambiguous, since no orientation of N has been specified.) The idea of the proof is to let ϵ approach zero.

Define
$$\begin{aligned} k : \mathrm{int}(V_\epsilon) &\longrightarrow SU(2) \\ X &\longmapsto X^{f(X)}. \end{aligned}$$

Note that
$$h_1(\rho)\alpha = \begin{cases} \rho(\alpha), & \alpha \in \pi_1(F_1) \\ k(\rho(\gamma))\rho(\alpha)k(\rho(\gamma))^{-1}, & \alpha \in \pi_1(F_2). \end{cases}$$

f can be chosen so that k is a diffeomorphism onto $SU(2) \setminus V_\epsilon$. Since $Q_1^\natural$ is (without loss of generality) transverse to $N^\natural$, there is a unique Ad-invariant diffeomorphism
$$l : (Q_1^\natural \cap N^\natural) \times V_\epsilon \to Q_1^\natural \cap N_\epsilon^\natural$$
such that for all $\rho \in Q_1^\natural \cap N^\natural$ and $X \in V_\epsilon$, $l(\rho, X)(\gamma) = X$.

It follows that as $\epsilon \to 0$, $h_1(Q_1^\natural \cap N_\epsilon^\natural)$ tends to the image of the map
$$m : (Q_1^\natural \cap N^\natural) \times (SU(2) \setminus \{-1\}) \longrightarrow N^\natural$$
$$m(\rho, X) = \begin{cases} \rho(\alpha), & \alpha \in \pi_1(F_1) \\ X\rho(\alpha)X^{-1}, & \alpha \in \pi_1(F_2), \end{cases}$$
while $Q_1^\natural \cap N_\epsilon^\natural$ tends to $Q_1^\natural \cap N^\natural = m(Q_1^\natural \cap N^\natural, -1)$. This implies (3.6). (The factor of 2 arises because Ad : $SU(2) \to \mathrm{Aut}(SU(2))$ has degree 2.) $\square$

(3.7) **Corollary.** *Let* $K \subset M$ *be a null-homologous knot in a* **Q***HS. Let* $K_{1/n}$ *denote* $1/n$ *Dehn surgery on* K. *Then there is a number* $\lambda'(K) \in \mathbf{Q}$ *such that*

$$\lambda(K_{1/n}) = \lambda(M) + n\lambda'(K)$$

for all n.

Proof: (Casson for **Z**HS case.) Let U be a regular neighborhood of K. Let $A \subset \partial U$ be a regular neighborhood of a simple closed curve representing the longitude of K. Then $1/n$ Dehn surgery on K is equivalent to cutting M along ∂U and regluing after performing n left-handed Dehn twists on A. The important thing to notice here is that this operation depends only on the annulus A and not on the cutting surface which contains it. Thus, by (3.2), it suffices to find a Heegaard surface for M which contains A as a separating annulus.

Let E be a Seifert surface for K. Fix an identification of $E \times I$ with a bicollar of E in M. Let C be a collar of ∂E in E. Then we can take U to be $C \times I$ and A to be $\partial U \times I$.

Pick a relative handle decomposition of $(M, E \times I)$. That is,

$$M = E \times I \cup \{1\text{-handles}\} \cup \{2\text{-handles}\} \cup \{3\text{-handles}\}.$$

After some sliding, we may assume that the attaching regions of the 1-handles lie in $E \times \{1\}$. It is now easy to see that $\partial(E \times I \cup \{1 - \text{handles}\})$ is the desired Heegaard surface. $\square$

(3.8) **Corollary.** *Let* $K, L \subset M$ *be a pair of null-homologous knots with linking number 0 in a* **Q***HS* M. *Let* $(K_{1/m}, L_{1/n})$ *denote the* **Q***HS obtained from* $1/m$ *Dehn surgery on* K *and* $1/n$ *Dehn surgery on* L. *Then there is a number* $\lambda''(K, L) \in \mathbf{Q}$ *such that*

$$\lambda(K_{1/m}, L_{1/n}) = \lambda(M) + m\lambda'(K) + n\lambda'(L) + mn\lambda''(K, L).$$

Proof: (Casson for **Z**HS case.) Note that since $\mathrm{lk}(K, L) = 0$, $1/n$ Dehn surgery on L does not affect the longitude of K, and vice-versa. The result now follows from (3.7). $\square$

Note that

$$(3.9) \quad \lambda''(K, L) \;=\; \lambda(K_{1/m}, L_{1/n}) - (K_{1/(m-1)}, L_{1/n})$$
$$-(K_{1/m}, L_{1/(n-1)}) + (K_{1/(m-1)}, L_{1/(n-1)})$$

for any $m, n \in \mathbf{Z}$.

(3.10) **Lemma.** *Let α and γ be disjoint separating simple closed curves on F. (It follows that α and γ are null-homologous in M and have linking number 0.) Then $\lambda''(\alpha, \gamma) = 0$.*

Proof: (Casson for **Z**HS case.) Let $h_\alpha\, [\, h_\gamma] : F \to F$ be a Dehn twist along $\alpha\, [\gamma]$. Let h_α and h_γ also denote the induced maps on $\pi_1(F)$ and R. Let

$$M_{m,n} \overset{\text{def}}{=} W_1 \cup_{h_\alpha^m h_\gamma^n} W_2.$$

This is the same as doing $1/m$ Dehn surgery on α and $1/n$ Dehn surgery on γ. By (3.9), it suffices to show that

$$\lambda(M_{1,1}) - \lambda(M_{1,0}) - \lambda(M_{0,1}) + \lambda(M_{0,0}) = 0$$

or, equivalently, that

$$\langle h_\alpha h_\gamma(Q_1), Q_2 \rangle - \langle h_\alpha(Q_1), Q_2 \rangle - \langle h_\gamma(Q_1), Q_2 \rangle + \langle Q_1, Q_2 \rangle = 0.$$

Let

$$N = \{[\rho] \in R \,|\, \rho(\gamma) = -1\} \subset R^-.$$

Note that h_α preserves N. Let $D \subset N$ be the difference cycle, as defined in the proof of (3.2), representing $h_\gamma(Q_1) - Q_1$. Then

$$\begin{aligned}
\langle h_\gamma(Q_1), Q_2 \rangle - \langle Q_1, Q_2 \rangle &= \langle D, Q_2 \rangle \\
\langle h_\alpha h_\gamma(Q_1), Q_2 \rangle - \langle h_\alpha(Q_1), Q_2 \rangle &= \langle h_\alpha(D), Q_2 \rangle.
\end{aligned}$$

h_α acts trivially on $H_*(F; \mathbf{Z})$ and so, by (1.16) and the universal coefficient theorem, also acts trivially on $H_{3g-3}(N; \mathbf{Q})$. Therefore

$$\langle h_\alpha(D), Q_2 \rangle = \langle D, Q_2 \rangle.$$

$\square$

(3.11) **Corollary.** *Let $K, L \subset M$ be null-homologous knots bounding disjoint Seifert surfaces. Then $\lambda''(K, L) = 0$.*

Proof: (Casson for **Z**HS case.) By (3.10), it suffices to situate K and L as disjoint separating curves on a Heegaard surface for M. This can be done as in the proof of (3.7). $\square$

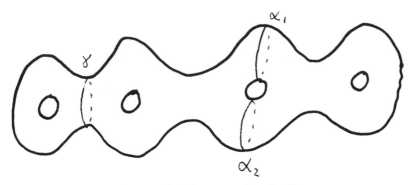

Figure 3.1: The setup for (3.12).

(3.12) **Lemma.** *Let* $\alpha_1, \alpha_2, \gamma \subset F$ *be disjoint simple closed curves such that* γ *separates* F *and* α_1 *and* α_2 *jointly separate* F *(see Figure 3.1). Let* $h_\alpha :$ $F \to F$ *be a right-handed Dehn twist along* α_1 *composed with a left-handed Dehn twist on* α_2. *Let* $M_n = W_1 \cup_{h_\alpha^n} W_2$. *Then* $\lambda'(\gamma; M_n) = \lambda'(\gamma; M_0)$ *for all* n, *where* $\lambda'(\gamma; M_n)$ *denotes* $\lambda'(\gamma)$ *with* γ *thought of as a knot in* M_n.

Proof: The proof is identical to that of (3.10), since the only properties of h_α used were that it preserved N_γ and acted trivially on $H^*(F; \mathbf{Z})$. $\quad\square$

(3.13) **Corollary.** *Let* $K, L_1, L_2 \subset M$ *be knots such that* K *bounds an embedded surface* E, L_1 *and* L_2 *cobound an embedded surface* H, *and* E *and* H *are disjoint. Let* M_n *be the result of performing* $1/n$ *Dehn surgery on* L_1 *and* $-1/n$ *Dehn surgery on* L_2, *where the surgery coefficients are given with respect to the framings determined by* H. *Then* $\lambda'(K; M_n) = \lambda'(K; M_0)$ *for all* n.

Proof: This follows from (3.12) and an argument similar to the one used in the proof of (3.7). $\quad\square$

The next lemma is most conveniently stated in terms of

(3.14) $$\bar{\lambda}(M) \stackrel{\text{def}}{=} |H_1(M; \mathbf{Z})| \lambda(M) = \langle Q_1, Q_2 \rangle.$$

The above results also hold with λ replaced with $\bar{\lambda}$. So we have, for example, an invariant $\bar{\lambda}'(K)$ for null-homologous knots K, and $\bar{\lambda}''(K, L) = 0$ if K and L are a boundary link.

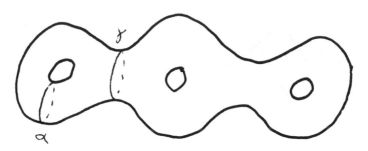

Figure 3.2: More curves called γ and α.

Let γ be a separating curve on F such that one component of $F \setminus \gamma$ has genus one. Let α be a non-separating curve in this component. Let h_α be a Dehn twist along α and let $M_n = W_1 \cup_{h_\alpha^n} W_2$ (see Figure 3.2). We wish to determine the effect of this operation on $\bar\lambda'(\gamma)$.

Modifying the gluing map of the Heegaard splitting by n Dehn twists along α is, of course, equivalent to performing $1/n$ Dehn surgery on α, with respect to the basis of $H_1(\partial(\mathrm{nbd}(\alpha)); \mathbf{Z})$ determined by the framing of α given by the normal bundle of α in F. Let p/q be the slope of the longitude of α with respect to this basis. After possibly changing the sign of one of the basis vectors and the sign of the Dehn twist, we may assume that $p/q < 0$.

(3.15) Lemma. $\bar\lambda'(\gamma; M_n) = \bar\lambda'(\gamma; M_0)$ *for all* $n > q/p$.

(If $n < q/p$, the statement remains true up to sign.)

Proof: Let h_γ be a Dehn twist along γ. Let h_α and h_γ also denote the induced maps on $\pi_1(F)$, R, etc. Let $N_\gamma = \{[\rho] \in R \mid \rho(\gamma) = -1\}$ and define N_α similarly. Let D be the difference cycle, as defined in (3.2), representing $h_\gamma(Q_1) - Q_1$. Thus we have

$$\bar\lambda'(\gamma; M_0) = \langle D, Q_2 \rangle.$$

Similarly,

$$\bar\lambda'(\gamma; M_n) = \langle h_\alpha^n(D), Q_2 \rangle.$$

Consider the action of the maps h_α^n ($n \in \mathbf{Z}$) on $T_1 R$. These maps are colinear in the vector space $\mathrm{End}(T_1 R)$ and so extend naturally to a one (real) parameter family of maps. It is easy to check that $h_\alpha^n(T_1 Q_1)$ is not transverse to $T_1 Q_2$ precisely when $n = q/p$.

Thus, for $n > q/p$ ($n \in \mathbf{Z}$), we can assign orientations (according to the conventions of (1.F)) so that $h^n_\alpha : Q_1 \to h^n_\alpha(Q_1)$ is orientation preserving . So, in order to show that $\bar{\lambda}'(\gamma; M_n) = \bar{\lambda}'(\gamma; M_0)$ (for $n > q/p$), it must be shown that

$$\langle h^n_\alpha(D), Q_2 \rangle = \langle D, Q_2 \rangle.$$

In fact, we will show that $h^n_\alpha(D)$ is isotopic to D.

Let F' be F cut along α. Let $\beta \subset F$ be a curve intersecting α once transversely. Then $\pi_1(F)$ is generated by $\pi_1(F')$ and β. If base points are chosen properly, the action of h_α on $\pi_1(F)$ is given by

$$\begin{aligned} h_\alpha(x) &= x, & x \in \pi_1(F') \\ h_\alpha(\beta) &= \alpha\beta. \end{aligned}$$

It follows that the action of h_α on $R^\natural$ is given by

$$\begin{aligned} h_\alpha(\rho)x &= \rho(x), & x \in \pi_1(F') \\ h_\alpha(\rho)\beta &= \rho(\alpha)\rho(\beta). \end{aligned}$$

Let $N_\alpha \overset{\text{def}}{=} \{[\rho] \in R \,|\, \rho(\alpha) = -1\}$. For $[\rho] \in R \setminus N_\alpha$ and $t \in \mathbf{R}$, define

$$\begin{aligned} h_t(\rho)x &= \rho(x), & x \in \pi_1(F') \\ h_t(\rho)\beta &= \rho(\alpha)^t \rho(\beta). \end{aligned}$$

(cf. (3.4).) For $n \in \mathbf{Z}$, $h^n_\alpha = h_n$. Hence h^n_α is isotopic to the identity on $R \setminus N_\alpha$.

Since γ bounds a genus one surface containing α as a non-separating curve, $\rho(\alpha) = -1$ implies $\rho(\gamma) = 1$. Since $D \subset N_\gamma = \{[\rho] \in R \,|\, \rho(\gamma) = -1\}$, this implies that $D \subset R^- \setminus N_\alpha$. Hence $h^n_\alpha(D)$ is isotopic to D. $\square$

(3.16) **Corollary.** *Let $K \subset M$ be a knot in a QHS bounding a genus one Seifert surface $E \subset M$. Let $L \subset E$ be a simple closed curve. Let $L_{1/n}$ denote $1/n$ Dehn surgery on L, with respect to the basis of $H_1(\partial(\mathrm{nbd}(L)); \mathbf{Z})$ determined by E (see the discussion preceding (3.15)). Let p/q be the slope of the longitude of L. Assume, without loss of generality, that $p/q < 0$. Then*

$$\bar{\lambda}'(K; L_{1/n}) = \bar{\lambda}'(K; M)$$

for all $n > q/p$.

Proof: Similar to the proof of (3.7). □

Let $K \subset M$ be a fibered knot in a **QHS**. Let E be a fiber and let $g : E \to E$ be the monodromy of the fibering. Let

$$Z \overset{\text{def}}{=} \text{Hom}(\pi_1(E), SU(2))/SU(2)$$

$$Z_{-1} \overset{\text{def}}{=} \{[\rho] \in Z \mid \rho(\partial E) = -1\}.$$

Note that Z_{-1} is a manifold.

Let $g^* : Z_{-1} \to Z_{-1}$ be the induced map. Let $\text{Lef}(g^*)$ denote the Lefschetz number of g^*. Recall that since Z_{-1} is a manifold, $\text{Lef}(g^*)$ is equal to the intersection number of the graph of g^* with the diagonal in $Z_{-1} \times Z_{-1}$.

(3.17) Lemma. *Let K and g^* be as above. Then $\bar{\lambda}'(K) = \pm 2 \text{Lef}(g^*)$.*

Proof: (Casson for **ZHS** case.) Let $p : M \setminus K \to S^1 = [0, 2\pi]/(0 \sim 2\pi)$ be the fibering projection. Let

$$E_0 \overset{\text{def}}{=} p^{-1}(0) \cup K$$

$$E_\pi \overset{\text{def}}{=} p^{-1}(\pi) \cup K$$

$$F \overset{\text{def}}{=} E_0 \cup E_\pi$$

$$W_1 \overset{\text{def}}{=} p^{-1}([0, \pi]) \cup K$$

$$W_2 \overset{\text{def}}{=} p^{-1}([\pi, 2\pi]) \cup K.$$

(F, W_1, W_2) is a Heegaard splitting of M. For the sake of notational consistency, let γ denote the image of K in F. We can take the fiber E above to be E_0. Identify E_0 and E_π via the product structure of W_1. With these identifications in mind, we have

$$
\begin{aligned}
R^\natural &= \{(\rho, \rho') \in Z^\natural \times Z^\natural \mid \rho(\gamma) = \rho'(\gamma)\} \\
Q_1^\natural &= \{(\rho, \rho') \in Z^\natural \times Z^\natural \mid \rho = \rho'\} \\
Q_2^\natural &= \{(\rho, \rho') \in Z^\natural \times Z^\natural \mid \rho = g^* \rho'\} \\
N^\natural &= \{(\rho, \rho') \in Z^\natural \times Z^\natural \mid \rho(\gamma) = \rho'(\gamma) = -1\},
\end{aligned}
$$

where N is as in the proof of (3.2).

There is a natural fibration

$$SO(3) \to N \overset{\pi}{\to} Z_{-1} \times Z_{-1}.$$

Let $\Delta \subset Z_{-1} \times Z_{-1}$ be the diagonal. Let $\tilde{\Delta} \stackrel{\text{def}}{=} \pi^{-1}(\Delta)$. Let D be the difference cycle, as defined in the proof of (3.2). Note that $\pi(Q_1 \cap N) = \Delta$. It follows from (3.6) that D is homologous to $2\tilde{\Delta}$.

The proof can now be completed using some standard intersection number yoga. We have (up to sign, with $\langle \cdot , \cdot \rangle_X$ denoting the intersection number in X),

$$
\begin{aligned}
\bar{\lambda}'(K) &= \langle D, Q_2 \rangle_R \\
&= 2\langle \tilde{\Delta}, Q_2 \rangle_R \\
&= 2\langle \tilde{\Delta}, Q_2 \cap N \rangle_N \\
&= 2\langle \Delta, \pi(Q_2 \cap N) \rangle_{Z_{-1} \times Z_{-1}} \\
&= 2\langle \Delta, \text{graph of } g^* \rangle_{Z_{-1} \times Z_{-1}} \\
&= 2\,\text{Lef}(g^*).
\end{aligned}
$$

$\square$

We are now in a position to do some calculations which will be needed in the proof of (3.36). Let $L_{p,q}$ denote the p, q lens space (i.e. p/q-surgery on the unknot in S^3). It is easy to see that, as oriented manifolds,

$$
L_{p,q} = -L_{p,p-q}.
$$

In particular, $L_{2,1} = -L_{2,1}$. Therefore, by (3.1),

(3.18) $$\bar{\lambda}(L_{2,1}) = 0.$$

Let K be a knot in S^3. Let $K^\dagger \subset S^3$ be a knot which differs from K by a crossing change (i.e. K and $K^\dagger$ coincide except in a 3-ball, where they differ as shown in Figure 3.3). Let $K_{p/q}$ denote p/q-surgery on K. Let $J \subset S^3 \setminus K$ be an unknot surrounding the crossing, as shown in Figure 3.4. Note that -1-surgery on J transforms K into $K^\dagger$. J bounds a genus one Seifert surface E consisting of an unknotting disk D for J with a neighborhood of $K \cap D$ removed and a tube surrounding half of K glued in. (See Figure 3.5.) Let $A \subset E$ be a curve isotopic in $S^3 \setminus K$ to a meridian on K. Note that in $K_{p/q}$, the longitude on A is a q/p curve.

Let $K_{p/q} J_{a/b} A_{c/d}$ denote the manifold obtained by doing p/q-surgery on K, a/b-surgery on J and c/d-surgery on A. By (3.16), -1-surgery on A in $K_{p/q}$ does not affect $\bar{\lambda}'(J)$, so long as $p/q > 0$. In other words,

$$
\begin{aligned}
\bar{\lambda}(K_{p/q} J_{-1/1} A_{-1/1}) - \bar{\lambda}(K_{p/q} J_{1/0} A_{-1/1}) = \\
\bar{\lambda}(K_{p/q} J_{-1/1} A_{1/0}) - \bar{\lambda}(K_{p/q} J_{1/0} A_{1/0}).
\end{aligned}
$$

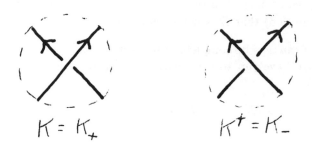

Figure 3.3: The non-coinciding parts of K and $K^\dagger$.

Figure 3.4: J, surgery upon which effects a crossing change

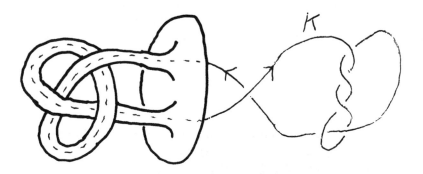

Figure 3.5: The genus one Seifert surface for J

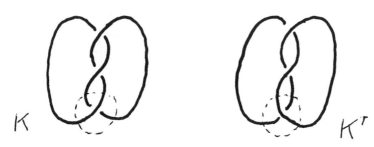

Figure 3.6: K and $K^\dagger$ (special case).

for all $p/q > 0$. This is equivalent to

$$(3.19) \qquad \bar\lambda(K^\dagger_{(p+q)/q}) - \bar\lambda(K^\dagger_{p/q}) = \bar\lambda(K_{(p+q)/q}) - \bar\lambda(K_{p/q}).$$

Now we specialize to the case where K is the right-handed trefoil knot and $K^\dagger$ is the unknot (see Figure 3.6). Four applications of (3.19) yield

$$(3.20) \qquad \bar\lambda(L_{5,1}) - \bar\lambda(L_{1,1}) = \bar\lambda(K_{5/1}) - \bar\lambda(K_{1/1}).$$

K is a fibered knot with genus one fiber (see, e.g., [Ro]). For the genus one case, it is not hard to see that Z_{-1} (defined in the proof of (3.17)) consists of a single point. Hence $\mathrm{Lef}(g^*) = \pm 1$ and, by (3.17), $\bar\lambda(K_{1/1}) = \bar\lambda'(K) = \lambda'(K) = \pm 2$. (It will follow from the proof of (3.36) that $\lambda'(K) = 2$.)

$L_{1,1} \cong S^3$, so $\bar\lambda(L_{1,1}) = 0$. It follows from [Mo] that $K_{5/1} \cong -L_{5,1}$ (see also Figure 3.7). Therefore, by (3.1) and (3.20),

$$2\bar\lambda(L_{5,1}) = \mp 2$$

or

$$(3.21) \qquad \bar\lambda(L_{5,1}) = \mp 1.$$

(3.22) **Lemma.** *Let M be a **Q**HS and Σ be a **Z**HS. Let K be a knot in Σ and let $K^\dagger$ be the corresponding knot in $\Sigma \# M$. Then*

$$\lambda'(K^\dagger) = \lambda'(K).$$

Proof: Let (W_{1a}, W_{2a}, F_a) $[(W_{1b}, W_{2b}, F_b)]$ be a Heegaard splitting of Σ $[M]$. Without loss of generality, K coincides a separating curve $\gamma \subset F_a$

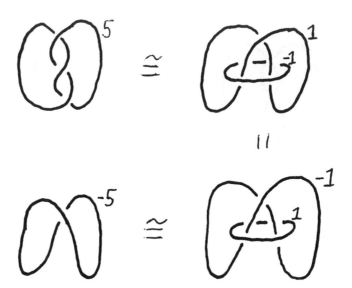

Figure 3.7: The proof that $K_{5/1} = -L_{5,1}$.

(see (3.7)). Let (W_1, W_2, F) be the connected sum of (W_{1a}, W_{2a}, F_a) and (W_{1b}, W_{2b}, F_b). (So that F is the connected sum of F_a and F_b, and W_j is the boundary connected sum of W_{ja} and W_{jb}. See II.2.c of [AM].) Let R, Q_j, S, T_j, etc. be as usual. Let E_x ($x = a$ or b) be the part of F_x coming from F. (So $E_x = F_x \setminus$ disk.) Then $F = E_a \cup E_b$ and $\alpha \overset{\text{def}}{=} \partial E_a = \partial E_b = E_a \cap E_b$.
    Let

$$
\begin{aligned}
R_x^{*\natural} &= \text{Hom}(\pi_1(E_x), SU(2)) \\
R_x^{\natural} &= \{\rho \in R_x^{*\natural} \mid \rho(\alpha) = 1\} \\
Q_{jx}^{\natural} &= \text{Hom}(\pi_1(W_{jx}), SU(2))
\end{aligned}
$$

($x = a$ or b). Then we can make the identification

$$
R^{\natural} = \{(\rho_a, \rho_b) \in R_a^{*\natural} \times R_b^{*\natural} \mid \rho_a(\alpha) = \rho_b(\alpha)\}.
$$

With respect to this identification,

$$
Q_j^{\natural} = Q_{ja}^{\natural} \times Q_{jb}^{\natural}.
$$

    Let $\{h_t^{\natural}\}$ $[\{h_{at}^{\natural}\}]$ be the isotopy of (3.2) for $\gamma \subset F$ $[\gamma \subset F_a]$, and let D $[D_a]$ be the associated difference cycle. Hence

$$
\lambda'(K^{\dagger}) = \frac{\langle D, Q_2 \rangle}{|H_1(M; \mathbf{Z})|}
$$

$$\lambda'(K) = \langle D_a, Q_{2a} \rangle.$$

To prove the lemma it suffices to show that

(3.23) $$\langle D, Q_2 \rangle = \pm |H_1(M; \mathbf{Z})| \langle D_a, Q_{2a} \rangle$$

and to check the sign.

It is easy to see that

$$h_t^\natural = h_{at}^\natural \times \mathrm{id}.$$

It follows that
(3.24) $$D^\natural = D_a^\natural \times Q_{1b}^\natural.$$

By choosing compatible isotopies on R_a^- and R^-, we may assume that D_a is in general position with respect to Q_{2a} and that (3.24) still holds.

Choose a neighborhood V_a of $D_a \cap Q_{2a}$ in R_a^- over which there is a smooth section $f : V_a \to R_a^\natural$ of the quotient map $R_a^\natural \to R_a$. Let U be a neighborhood of 1 in $SU(2)$. Choose an immersion

$$\bar{f} : V_a \times U \to R_a^{*\natural}$$

so that $\bar{f}(\,\cdot\,, 1) = f$ and $\bar{f}(p, u)(\alpha) = u$. Let $N_b^\natural$ be a neighborhood of $R_b^\natural$ in $R_b^{*\natural}$.

Define

$$\begin{aligned} g : V_a \times V_b^\natural &\to R^- \\ (p_a, \rho_b) &\mapsto \pi(\bar{f}(p_a, \rho_b(\alpha)), \rho_b), \end{aligned}$$

where π is the quotient map from $R^\natural \subset R_a^{*\natural} \times R_b^{*\natural}$ to R. It is easy to see that g is a diffeomorphism onto its image. Furthermore,

$$\begin{aligned} D \cap Q_2 &\subset g(V_a \times V_b^\natural) \\ g^{-1}(D) &= (V_a \cap D_a) \times Q_{1b}^\natural \\ g^{-1}(Q_2) &= (V_a \cap Q_{2a}) \times Q_{2b}^\natural. \end{aligned}$$

It follows that

$$\langle D, Q_2 \rangle = \pm \langle D_a, Q_{2a} \rangle \langle Q_{1b}^\natural, Q_{2b}^\natural \rangle_{R_b^{*\natural}}.$$

But by Proposition III.1.1 of [AM],

$$\langle Q_{1b}^\natural, Q_{2b}^\natural \rangle_{R_b^{*\natural}} = \pm |H_1(M; \mathbf{Z})|.$$

This proves (3.23).

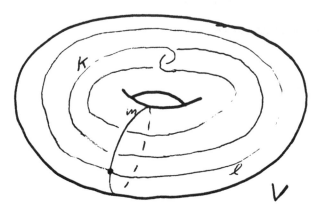

Figure 3.8: A pre-untwisted double.

To determine the sign, one need only consider orientations near the trivial representation. Since a neighborhood of the trivial representation is independent (up to diffeomorphism) of the manifolds involved, it suffices to check a single example, e.g. $M = S^3$. The lemma is clearly true in this case, so it is true in general. $\square$

In the proof of the next lemma we will need the concept of a clean intersection. Let A and B be oriented submanifolds of an oriented manifold M which intersect in an (oriented) submanifold Z. Assume that $\dim(A) + \dim(B) = \dim(M)$. (Note that A and B are not transverse if $\dim(Z) > 0$.) A and B are said to intersect *cleanly* if

$$TZ = TA|_Z \cap TB|_Z.$$

Let $\nu(Z)$ be the intersection of the normal bundles of A and B in M, restricted to Z. $\nu(Z)$ is an orientable, $\dim(Z)$ dimensional vector bundle over Z, and it is not hard to see that

(3.25) $\langle A, B \rangle = \pm e(\nu(Z)),$

where $e(\nu(Z))$ denotes the Euler number of $\nu(Z)$.

Consider the solid torus V containing the knot K, as shown in Figure 3.8. If V is embedded is a QHS M in such a way that the curve $l \in \partial V$ is parallel to a longitude of $\overline{M \setminus V}$, then we say that K (considered as a knot in M) is an *untwisted double* (with negative clasp).

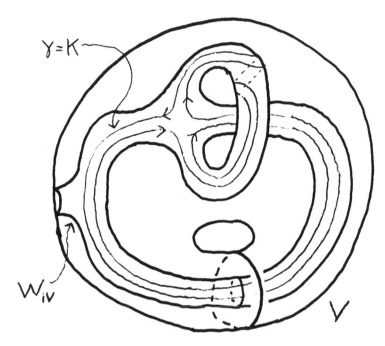

Figure 3.9: $W_{1V} \subset V$

(3.26) **Lemma.** *Let* $K \subset M$ *be an untwisted double. Then* $\lambda'(K) = 0$.

Proof: As usual, we begin by constructing the appropriate Heegaard splitting. Fix some embedding of V in M which realizes K as an untwisted double. Identify V with its image under this embedding. K lies on the surface of a genus two handlebody $W_{1V} \subset V$, as shown in Figure 3.9. (W_{1V} can be thought of as ⟨a regular neighborhood of the core of V⟩ union ⟨an unknotted circle which intersects the core once⟩ union ⟨an arc connecting the core to ∂V⟩.) Let W_{2V} be the closure in V of $V \setminus W_{1V}$. Let $\Lambda \overset{\text{def}}{=} \overline{M \setminus V}$. Choose a collar $\partial V \times I \subset \Lambda$ of ∂V in Λ. Let $W_{2\Lambda} \subset \Lambda$ be $(W_{2V} \cap \partial V) \times I$ union some 1-handles in Λ. Let $W_{1\Lambda}$ be the closure in Λ of $\Lambda \setminus W_{2\Lambda}$. If the 1-handles are chosen properly, then $W_{1\Lambda}$ will be a handlebody, and $W_1 \overset{\text{def}}{=} W_{1V} \cup W_{1\Lambda}$ and $W_2 \overset{\text{def}}{=} W_{2V} \cup W_{2\Lambda}$ will form a Heegaard splitting of M. Let $F \overset{\text{def}}{=} \partial W_1 = \partial W_2$. Let $F_V \overset{\text{def}}{=} F \cap V$ and $F_\Lambda \overset{\text{def}}{=} F \cap \Lambda$. Note that the natural maps $\pi_1(F_V) \to \pi_1(W_{jV})$ and $\pi_1(F_\Lambda) \to \pi_1(W_{j\Lambda})$ $(j = 1, 2)$ are all surjective. For the sake of notational consistency, let γ denote K thought of as a curve in F.

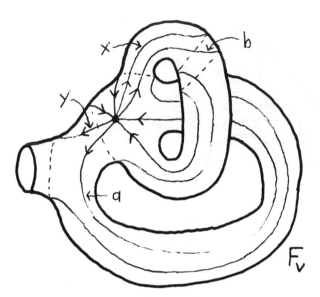

Figure 3.10: a, b, x and y.

Let R, Q_j, S, etc., be as usual. Let

$$R_V^{*\natural} \quad \overset{\text{def}}{=} \quad \text{Hom}(\pi_1(F_V), SU(2))$$

$$Q_{jV}^\natural \quad \overset{\text{def}}{=} \quad \text{Hom}(\pi_1(W_{jV}), SU(2)).$$

Define $R_\Lambda^{*\natural}$ and $Q_{j\Lambda}^\natural$ similarly. $R^\natural$ can be identified with a subset of $R_V^{*\natural} \times R_\Lambda^{*\natural}$. It is easy to see that with respect to this identification, $Q_1^\natural = Q_{1V}^\natural \times Q_{1\Lambda}^\natural$ and $Q_2^\natural \subset Q_{2V}^\natural \times Q_{2\Lambda}^\natural$. ($Q_2^\natural$ is equal to the set of all $(\rho_V, \rho_\Lambda) \subset Q_{2V}^\natural \times Q_{2\Lambda}^\natural$ which agree on $\pi_1(W_2 \cap \partial V)$.)

Choose a set of generators $a, b, x, y \in \pi_1(F_V)$ as shown in Figure 3.10. For $\rho \in R_V^{*\natural}$, let $A \overset{\text{def}}{=} \rho(a)$, $B \overset{\text{def}}{=} \rho(b)$, $X \overset{\text{def}}{=} \rho(x)$, $Y \overset{\text{def}}{=} \rho(y)$. The functions A, B, X and Y give an identification of $R_V^{*\natural}$ with $SU(2)^4$.

Let $c, d, e, \partial \in \pi_1(F_V)$ be as shown in Figure 3.11. We have

$$
\begin{aligned}
c &= xy^{-1}x^{-1}bab^{-1} \\
d &= x^{-1}b \\
e &= ba^{-1}b^{-1}xyx^{-1}b \\
\partial &= aba^{-1}b^{-1}xyx^{-1}y^{-1} \\
\gamma &= aba^{-1}b^{-1}.
\end{aligned}
$$

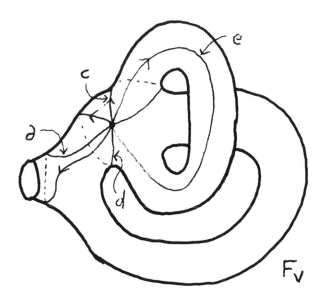

Figure 3.11: c, d, e and ∂

(In the last equation, γ is confused with an element of $\pi_1(F_V)$ in its conjugacy class.) It follows that

$$
\begin{aligned}
Q^\natural_{1V} &= \{(A,B,X,Y) \in R^{*\natural}_V \mid XY^{-1}X^{-1}BAB^{-1} = 1, X^{-1}B = 1\} \\
&= \{(A,B,X,Y) \in R^{*\natural}_V \mid X = B, Y = A\} \\
Q^\natural_{2V} &= \{(A,B,X,Y) \in R^{*\natural}_V \mid BA^{-1}B^{-1}XYX^{-1}B = 1\}.
\end{aligned}
$$

Let $D \subset R$ be the difference cycle associated to γ, as described in the proof of (3.2). The lemma will be proved if it can be shown that $\langle D, Q_2 \rangle = 0$.

Our next goal is to describe D more explicitly. Let F_1 and F_2 be the components of F cut along γ. $R^\natural$ can be identified with a subset of $\mathrm{Hom}(\pi_1(F_1), SU(2)) \times \mathrm{Hom}(\pi_1(F_2), SU(2))$. Let $N^\natural \overset{\mathrm{def}}{=} \{\rho \in R^\natural \mid \rho(\gamma) = -1\}$. It is not hard to show that Q_1 intersects N transversely. By (3.6) we may assume

(3.27) $D^\natural = 2\{(\rho_1, g\rho_2) \in R^\natural \mid g \in \mathrm{Aut}\,(SU(2)), (\rho_1, \rho_2) \in N^\natural \cap Q^\natural_1\}$,

where $(g\rho_2)(z) \overset{\mathrm{def}}{=} g(\rho_2(z))$.

In the present situation,

$$
N^\natural \cap Q^\natural_1 = \{(A,B,X,Y) \in R^\natural_V \mid ABA^{-1}B^{-1} = -1, X = B, Y = A\} \times Q^\natural_{1\Lambda}.
$$

(Here we are thinking of $R^\natural$ as a subset of $R_V^{*\natural} \times R_\Lambda^{*\natural}$ again.) γ is isotopic to the boundary of a regular neighborhood of curves representing a and b. It follows from (3.27) that we take $D^\natural$ to be $D_V^\natural \times Q_{1\Lambda}^\natural$, where

$$
\begin{aligned}
D_V^\natural \ &= \ 2\{(gA, gB, X, Y) \in R_V^\natural \,|\, g \in \mathrm{Aut}\,(SU(2)), \\
& \qquad\qquad\qquad ABA^{-1}B^{-1} = -1, X = B, Y = A\} \\
&= \ 2\{(gA, gB, X, Y) \in R_V^\natural \,|\, ABA^{-1}B^{-1} = -1, XYX^{-1}Y^{-1} = -1\}.
\end{aligned}
$$

Let $U \in \mathbf{su}(2)$ be a ball around 0 such that $\exp : U \to SU(2) \setminus \{-1\}$ is a diffeomorphism. Let $L \subset U$ be the 2-sphere which maps to the elements of trace 0. Scale the metric on $\mathbf{su}(2)$ so that L is the sphere of radius $\pi/2$. Then $ABA^{-1}B^{-1} = -1$ if and only if $A = \exp(u)$ and $B = \exp(v)$ where $u, v \in L$ and u is orthogonal to v.

Let $R_V^\natural \overset{\mathrm{def}}{=} \{\rho \in R_V^{*\natural} \,|\, \rho(\partial) = 1\}$. Let $S_V^\natural$ be the abelian representations in $R_V^\natural$. Define $R_\Lambda^\natural$ and $S_\Lambda^\natural$ similarly. Let $Z^\natural \overset{\mathrm{def}}{=} Z_V^\natural \times R_\Lambda^\natural$, where

$$
Z_V^\natural \overset{\mathrm{def}}{=} \{\rho \in R_V^\natural \,|\, \rho(y) \neq \pm 1\}.
$$

For $\rho \in Z_V^\natural$, let $u(\rho) \in U$ be the unit vector such that

$$
\exp(tu(\rho)) = \rho(y)
$$

for some $0 < t < \pi$. There is an S^1-action on $Z_V^\natural$ defined by

$$
(\theta\rho)(z) = \exp(\theta u(\rho))\rho(z)\exp(-\theta u(\rho))
$$

$(\rho \in Z_V^\natural, z \in \pi_1(F_V), \theta \in [0, \pi]/(0 \sim \pi) = S^1)$. By acting trivially on the second factor, this induces an action of S^1 on $Z_V^\natural \times R_\Lambda^\natural = Z^\natural$. This action clearly descends to one on $Z = Z^\natural / SU(2)$. S^1 acts freely on

$$
(3.28) \quad Z_- \overset{\mathrm{def}}{=} Z \setminus \left((Z_V^\natural \times S_\Lambda^\natural)/SU(2) \,\cup\, ((S_V^\natural \cap Z_V^\natural) \times R_\Lambda^\natural)/SU(2)\right).
$$

Note that both $D = D \cap Z$ and $Q_2 \cap Z$ are invariant under this action. Thus if we can equivariantly isotope D in Z so that it intersects Q_2 in isolated principal orbits and these intersections are clean, then it will follow from (3.25) that $\langle D, Q_2 \rangle$ is equal to the Euler number of an oriented vector bundle over a closed 1-manifold, and hence is zero.

The first step will be to isotope D in Z so that $D \cap Z \subset Z_-$.

Let $\rho_V = (A, B, X, Y) \in D_V^\natural$. We will define a deformation $\rho_{V,t} = (A_t, B_t, X_t, Y_t)$ $(t \in [0, \epsilon], \epsilon$ near 0). Define

$$
\begin{aligned}
X_t \ &= \ X \\
(3.29) \qquad\qquad Y_t \ &= \ \exp(tu(\rho_V))Y.
\end{aligned}
$$

It follows that
$$X_t Y_t X_t^{-1} Y_t^{-1} = -\exp(-2tu(\rho_V)).$$

Assume for the moment that we can find A_t and B_t (depending smoothly on ρ_V and t, and such that $A_0 = A$ and $B_0 = B$) which satisfy

(3.30) $$A_t B_t A_t^{-1} B_t^{-1} = -\exp(2tu(\rho_V))$$

and

(3.31) $(\mathrm{Ad}\,(H)\rho_V)_t = (\mathrm{Ad}\,(H)A_t, \mathrm{Ad}\,(H)B_t, \mathrm{Ad}\,(H)X_t, \mathrm{Ad}\,(H)Y_t)$

for all $H \in SU(2)$. It follows that
$$\rho_{V,t}(\partial) = A_t B_t A_t^{-1} B_t^{-1} X_t Y_t X_t^{-1} Y_t^{-1} = 1,$$

and hence that $\rho_{V,t} \in Z_V^{\natural}$.

Let $\{D_{V,t}^{\natural}\}$ denote the resulting deformation of $D_V^{\natural}$. Define a deformation $\{D_t^{\natural}\}$ of $D^{\natural}$ by $(\rho_V, \rho_\Lambda)_t = (\rho_{V,t}, \rho_\Lambda)$ for $(\rho_V, \rho_\Lambda) \in D_V^{\natural} \times Q_{1\Lambda}^{\natural} = D^{\natural}$. It follows from (3.31) that this deformation descends to a deformation $\{D_t\}$ of D in Z which commutes with the action of S^1 on Z.

Now to define A_t and B_t. Consider the map

$$\begin{aligned} c : SU(2) \times SU(2) &\longrightarrow SU(2) \\ (G, H) &\longmapsto GHG^{-1}H^{-1}. \end{aligned}$$

Let $W \overset{\text{def}}{=} SU(2) \setminus \{1\}$. c is a submersion, and hence a fibering, over W (with fiber isomorphic to $c^{-1}(-1) \cong SO(3)$). Suppose that there is a smooth bundle map

$$f : W \times c^{-1}(-1) \to c^{-1}(W)$$

such that $f(-1, x) = x$ and such that f commutes with the actions of $SU(2)$ on $W \times c^{-1}(-1)$ and $c^{-1}(W)$. Then

$$(A_t, B_t) \overset{\text{def}}{=} f(-\exp(2tu(\rho_V)), (A, B))$$

clearly satisfies (3.30) and (3.31).

To define f, choose an $SU(2)$-invariant connection on the fibering

$$c^{-1}(W) \to W.$$

Let $f(w, x)$ (for $w \neq -1$) be the parallel transport of x along the arc connecting -1 to w in the abelian subgroup containing w. It is easy to see that this works.

Next we show that for some (in fact almost all) $t \in [0, \epsilon]$, $D_t \cap Q_2 \subset Z_-$. Clearly $D^\natural_{V,t} \cap S^\natural_V = \emptyset$ for all t. So, by (3.28), what remains to be shown is that

(3.32) $$(D^\natural_t \cap Q^\natural_2) \cap (Z^\natural_V \times S^\natural_\Lambda) = \emptyset.$$

Note that

$$D^\natural_t \cap Q^\natural_2 \cap (Z^\natural_V \times S^\natural_\Lambda) \subset ((D^\natural_{V,t} \times Q^\natural_{1\Lambda}) \cap (Q^\natural_{2V} \times Q^\natural_{2\Lambda})) \cap (Z^\natural_V \times S^\natural_\Lambda)$$
$$= (D^\natural_{V,t} \cap Q^\natural_{2V}) \times (Q^\natural_{1\Lambda} \cap Q^\natural_{2\Lambda} \cap S^\natural_\Lambda).$$

$Q^\natural_{1\Lambda} \cap Q^\natural_{2\Lambda} \cap S^\natural_\Lambda$ is just the space of abelian representations $\pi_1(\Lambda) \to SU(2)$. Since K is an *untwisted* double, the curve l in Figure 3.8 must have finite order (say $d \in \mathbf{Z}$) in $H_1(\Lambda; \mathbf{Z})$. Thus if $\bar{l} \in \pi_1(F_V)$ is freely homotopic in Λ to l and $\rho_\Lambda \in Q^\natural_{1\Lambda} \cap Q^\natural_{2\Lambda} \cap S^\natural_\Lambda$, then $\rho_\Lambda(\bar{l})^d = 1$. On the other hand, l is freely homotopic in W_{2V} to y (see Figures 3.9 and 3.10). Therefore if

$$\rho \in ((D^\natural_{V,t} \cap Q^\natural_{2V}) \times (Q^\natural_{1\Lambda} \cap Q^\natural_{2\Lambda} \cap S^\natural_\Lambda)) \cap Q^\natural_2,$$

then $Y^d_t = \rho(y)^d = 1$. It is clear from (3.29) that this is possible for only a finite number of t. Thus (3.32) holds for all but finitely many t. Let t_0 be one of these and define $D' \overset{\text{def}}{=} D_{t_0}$. D' is the desired deformation of D which contains only principal orbits.

Now for the clean intersection property. First we show that near $D' \cap Q_2$, $Q_2 \cap Z$ is a manifold of dimension $3g - 5$ (where g is the genus of F). There is a free basis $\bar{m}, \bar{l}, x_3, \ldots, x_g$ of $\pi_1(W_2)$ such that $\bar{m}$ [$\bar{l}$] is freely homotopic to m [l] and such that ∂ is freely homotopic to $\bar{m}\bar{l}\bar{m}^{-1}\bar{l}^{-1}$. Hence $Q_2 \cap Z$ is diffeomorphic to

$$\{(\rho(\bar{m}), \rho(\bar{l}), \rho(x_3), \ldots, \rho(x_g)) \in SU(2)^g \mid$$
$$\rho(\bar{m})\rho(\bar{l})\rho(\bar{m})^{-1}\rho(\bar{l})^{-1} = 1, \rho(\bar{l}) \neq \pm 1\}/SU(2).$$

It is now easy to see that the irreducible part of $Q_2 \cap Z$ is a manifold of dimension $3g - 5$. Since D' consists entirely of irreducible representations, we are done.

Let $\nu(Q_2 \cap Z; Q_2)$ denote the normal bundle of $Q_2 \cap Z$ in Q_2. Let $\nu(Z; R)$ denote the normal bundle of Z in R. I claim that near $D' \cap Q_2$

(3.33) $$\nu(Q_2 \cap Z; Q_2) = TQ_2 \cap \nu(Z; R).$$

Let $f : R^\natural \to SU(2)$ be given by $f(\rho) \overset{\text{def}}{=} \rho(\partial)$. (3.33) follows from the facts that $Z^\natural$ is an open submanifold of $f^{-1}(1)$, f is a submersion near $D'^\natural \cap Q^\natural_2$, and $f|_{Q^\natural_2}$ has rank 2.

Now isotope D'/S^1 inside Z/S^1 so that it is in general position with respect to $(Q_2 \cap Z)/S^1$ and so that $(D'/S^1) \cap ((Q_2 \cap Z)/S^1) \subset Z_-/S^1$. Cover this isotopy with an isotopy of D'. $D' \cap Q_2$ is now a closed 1-manifold. (3.33) and the fact that D'/S^1 is in general position with respect to $(Q_2 \cap Z)/S^1$ imply that D' and Q_2 intersect cleanly. Hence $\langle D', Q_2 \rangle$ is equal to the Euler number of an oriented line bundle over a closed 1-manifold, i.e. $\langle D', Q_2 \rangle = 0$. $\qquad\square$

(3.34) **Lemma.** *Let K be a knot in a $\mathbf{Q}HS$ M with self-linking zero. That is, there exists $x \in H_1(\partial(\mathrm{nbd}(K)); \mathbf{Z})$ such that $\langle m, x \rangle = 1$ and $l = \langle m, l \rangle x$, where m is the meridian and l is the longitude of K. Let $K_{1/n}$ denote $1/n$-surgery on K, with respect to the basis m, x. Then there exists $\lambda'(K) \in \mathbf{Q}$ such that*

$$\lambda(K_{1/n}) = \lambda(M) + n\lambda'(K).$$

Note that if $\langle m, x \rangle = \pm 1$, this is (3.7). Note also that since $|H_1(K_{1/n}; \mathbf{Z})| = |H_1(M; \mathbf{Z})|$, $\bar{\lambda}(K_{1/n}) = \bar{\lambda}(M) + n\bar{\lambda}'(K)$, where $\bar{\lambda}'(K) \stackrel{\text{def}}{=} |H_1(M; \mathbf{Z})|\lambda'(K)$.

Proof: First we construct the appropriate Heegaard splitting of M. Let U be a closed regular neighborhood of K. Let $d \stackrel{\text{def}}{=} |\langle m, l \rangle|$. Assume that $d > 1$. Let $E \subset \overline{M \setminus U}$ be a Seifert surface for K. That is, $[\partial E] = l \in H_1(\partial U; \mathbf{Z})$ and ∂E consists of d disjoint parallel curves. Let $E \times [-1, 1] \subset M$ be a bicollar of E. As in the proof of (3.7), we can add 1-handles to $E \times \{-1\}$ so that $W_2 \stackrel{\text{def}}{=} E \times [-1, 1] \cup \{1 - \text{handles}\}$ and $W_1 \stackrel{\text{def}}{=} \overline{M \setminus W_2}$ are a Heegaard splitting of M. As usual, let $F \stackrel{\text{def}}{=} \partial W_1 = \partial W_2$.

Let $\gamma_1, \dots, \gamma_d$ be the boundary components of $E = E \times \{0\}$, thought of as curves on F. Let h be a left-handed Dehn twist along γ_1. Then

$$K_{1/n} = W_1 \cup_{h^n} W_2,$$

so what must be shown is that

$$\langle h^n(Q_1), Q_2 \rangle - \langle h^{n-1}(Q_1), Q_2 \rangle$$

is independent of n. The proof of this is similar to that of (3.2). Extra complications arise because γ_1 is non-separating.

Let $a \subset E$ be an arc connecting γ_1 to, say, γ_2. Let $\beta \subset F$ be the loop $\partial a \times [-1, 1] \cup a \times \{1\} \cup a \times \{-1\}$. Note that β and γ_1 intersect once transversely. Let $F' \subset F$ be F cut open along γ_1. Then the action of h on $\pi_1(F)$ and $R^\sharp$ is as described in the proof of (3.15) (with h replacing h_α and γ_1 replacing α).

Let $V_\epsilon \subset SU(2)$ be the closed ball of radius ϵ (with respect to some Ad-invariant metric) centered at $-1 \in SU(2)$. Let

$$
\begin{aligned}
N^\natural &= \{\rho \in R^\natural \,|\, \rho(\gamma_1) = -1\} \\
N^\natural_\epsilon &= \{\rho \in R^\natural \,|\, \rho(\gamma_1) \in V_\epsilon\}.
\end{aligned}
$$

Let $f : SU(2) \to [0,1]$ be a smooth, Ad-invariant, monotonic (on $SU(2)/\mathrm{Ad}$) function such that $f = 1$ outside V_ϵ and $f(-1) = 0$. For $t \in [0,1]$, define $h^\natural_t : R^\natural \to R^\natural$ by

$$
\begin{aligned}
h^\natural_t(\rho)x &= \rho(x), & x \in \pi_1(F') \\
h^\natural_t(\rho)\beta &= \rho(\gamma_1)^{f(\rho(\gamma_1))t}\rho(\beta).
\end{aligned}
$$

$\{h^\natural_t\}$ descends to an isotopy $\{h_t\}$ of R. Note that $h_0 = \mathrm{id}$ and $h_1 = h$ on $R \setminus N_\epsilon$.

Let $[\rho] \in T_1$ (i.e. ρ is an abelian representation of $\pi_1(W_1)$). Since the γ_i's are all homologous in W_1, $\rho(\gamma_1) = \rho(\gamma_2) = \cdots = \rho(\gamma_d)$. Since $\gamma_1 \cup \cdots \cup \gamma_d$ is a boundary in W_1, $\rho(\gamma_1)^d = 1$. Since T_1 is connected and $\rho(\gamma_1) = 1$ if ρ is the trivial representation, $\rho(\gamma_1) = 1$ for all $[\rho] \in T_1$. Therefore $h_t|_{T_1} = \mathrm{id}$.

It follows from Theorem 4.7 of [G3] that h_t is a symplectic map of R. Hence h_t acts symplectically on ν. It is easy to see that $h_t(Q_1)$ is transverse to Q_2 at P. $h_t|_S$ is not the identity, so $\{h_t\}$ is not quite a special isotopy. However, since $h_t|_{T_1} = \mathrm{id}$, the proof of special isotopy invariance in Section 2 goes through anyway and we have

$$
\langle Q_1, Q_2 \rangle = \langle h_1(Q_1), Q_2 \rangle.
$$

Let

$$
D \overset{\mathrm{def}}{=} h(Q_1 \cap N_\epsilon) \cup (-h_1(Q_1) \cap N_\epsilon).
$$

Since $T_1 \cap N = \emptyset$, $D \subset R^-$. Therefore

$$
\langle h(Q_1), Q_2 \rangle - \langle Q_1, Q_2 \rangle = \langle D, Q_2 \rangle,
$$

where the right hand side denotes a homological intersection number.

As in the proof of (3.2), it now suffices to show that $h(D)$ is homologous to D. This will follow if it can be shown that h acts trivially on $H_*(N; \mathbf{Q})$, since D can be homotoped into N.

Let $F^\dagger \subset F$ be the complement of a regular neighborhood of $\gamma_1 \cup \beta$. Let

$$
R^{\dagger\natural} \overset{\mathrm{def}}{=} \{\rho \in \mathrm{Hom}(\pi_1(F^\dagger), SU(2)) \,|\, \rho(\partial F^\dagger) = 1\}.
$$

Define $S^{\dagger\natural}$ and $P^{\dagger\natural}$ similarly. $N^\natural$ can be identified with $R^{\dagger\natural} \times SU(2)$, where the second factor is identified with $\rho(\beta)$. With respect to this identification,

$$
h(\tau, X) = (\tau, -X).
$$

Note that $h^2 = $ id.

There is a natural map

$$\pi : N \to R^{\dagger}.$$

$\pi^{-1}(p)$ is diffeomorphic to $SU(2)$, D^2 or I, according to whether p is in $R^{\dagger -}$, $S^{\dagger -}$ or $P^{\dagger}$, respectively.

We will need the following fact, the proof of which is left to the reader.

(3.35) *Let f be an endomorphism of a long exact sequence of vector spaces over* **Q**

$$\cdots \to A_i \to B_i \to C_i \to A_{i-1} \to \cdots$$

such that $f^2 = $ id, $f|_{A_i} = $ id and $f|_{B_i} = $ id (for all i). Then $f|_{C_i} = $ id for all i.

h descends to the identity on $R^{\dagger}$. Since $\pi : \pi^{-1}(P^{\dagger}) \to P^{\dagger}$ is a homotopy equivalence, h acts trivially on $H_*(\pi^{-1}(P^{\dagger}); \mathbf{Q})$. Similarly, h acts trivially on $H_*(\pi^{-1}(S^{\dagger}_-); \mathbf{Q})$. Applying (3.35) to the Gysin sequence of the fibering $\pi^{-1}(R^{\dagger}_-) \to R^{\dagger}_-$ we see that h acts trivially on $H_*(\pi^{-1}(R^{\dagger}_-); \mathbf{Q})$. Let U be regular neighborhood of $P^{\dagger}$ in $S^{\dagger}$. Applying (3.35) to the Mayer-Vietoris sequence of the triple $(\pi^{-1}(S^{\dagger}), \pi^{-1}(S^{\dagger}_-), \pi^{-1}(U))$ yields that h acts trivially on $H_*(\pi^{-1}(S^{\dagger}); \mathbf{Q})$. (This uses the easily proved fact that h acts trivially on $H_*(\pi^{-1}(S^{\dagger}_- \cap U); \mathbf{Q})$.) Similarly, h acts trivially on $H_*(\pi^{-1}(R^{\dagger}); \mathbf{Q}) = H_*(N; \mathbf{Q})$.   □

(3.36) **Lemma.** *Let $K \subset M$ be a knot in a* **Q***HS and let $N \overset{\text{def}}{=} M \setminus \text{nbd}(K)$. Let $l \in H_1(\partial N; \mathbf{Z})$ be a longitude of K and let $a, b \in H_1(\partial N; \mathbf{Z})$ be primitive homology classes such that $\langle a, l \rangle \langle b, l \rangle > 0$ and $\langle a, b \rangle = 1$. Let $c, d \in H_1(\partial N; \mathbf{Z})$ be primitive homology classes such that $\langle a, c \rangle \geq 0$, $\langle a, d \rangle \geq 0$, $\langle c, b \rangle \geq 0$, $\langle d, b \rangle \geq 0$ and $\langle c, d \rangle = 1$ (see Figure 3.12). For primitive $x \in H_1(\partial N; \mathbf{Z})$, let N_x denote Dehn surgery along x. Then there is a real number $r(K, a, b)$ (depending on K, a and b, but not on c and d) such that*

$$\bar{\lambda}(N_{c+d}) - \bar{\lambda}(N_c) - \bar{\lambda}(N_d) - \frac{1}{6}(|H_1(N_c; \mathbf{Z})| - |H_1(N_d; \mathbf{Z})|) = r(K, a, b).$$

Remarks: It will be shown in the next section that $r(K, a, b) = 0$ for all K, a and b. Note that (in view of the lemma) knowledge of $\bar{\lambda}(N_a)$, $\bar{\lambda}(N_b)$ and $r(K, a, b)$ determines $\bar{\lambda}(N_c)$ for all primitive $c \in H_1(\partial N; \mathbf{Z})$ lying between a and b (i.e. $\langle a, c \rangle \geq 0$, $\langle c, b \rangle \geq 0$).

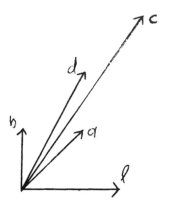

Figure 3.12: a, b, c, d and l.

Proof: The proof is long and technical.

For a, b as in the statement of the lemma, define

$$r(a,b) \overset{\text{def}}{=} \bar{\lambda}(N_{a+b}) - \bar{\lambda}(N_a) - \bar{\lambda}(N_b) - \frac{1}{6}(|H_1(N_a; \mathbf{Z})| - |H_1(N_b; \mathbf{Z})|).$$

The lemma will be proved if it can be shown that

$$\begin{aligned} r(a, a+b) &= r(a,b) \\ r(a+b, b) &= r(a,b), \end{aligned}$$

for any pair (c, d) (as in the statement of the lemma) is related to (a, b) by a sequence of the elementary transformations $a \mapsto a$, $b \mapsto a+b$ and $a \mapsto a+b$, $b \mapsto b$.

First we construct the appropriate Heegaard splitting. Let $U \overset{\text{def}}{=} \text{nbd}(K)$. Let α and β be simple closed curves on ∂U $(= \partial N)$ which represent a and b and which intersect once transversely. Attach 1-handles to U, disjoint from α and β, so that $W_1 \overset{\text{def}}{=} U \cup \{\text{1-handles}\}$ and $W_2 \overset{\text{def}}{=} \overline{M \setminus W_1}$ form a Heegaard splitting of M. Let F, R, Q_j, etc. be as usual. Let $\gamma \subset F$ be the boundary of a regular neighborhood of $\alpha \cup \beta$ in F. Let $\delta_1, \ldots, \delta_{g-1} \subset F$ be the boundaries of disks transverse to the cores of the 1-handles. (Note that $\rho(\delta_i) = 1$ for $[\rho] \in Q_1$.)

Let $z \in H_1(\partial U; \mathbf{Z})$ be primitive and let $\xi \subset \partial U$ be a simple closed curve representing z and disjoint from γ. (Disjointness from γ guarantees that $\xi \subset F$.) Define

$$\begin{aligned} Q_1^z &\overset{\text{def}}{=} \{[\rho] \in R \mid \rho(\xi) = 1, \rho(\delta_i) = 1, 1 \le i \le g-1\} \\ Q_1^{-z} &\overset{\text{def}}{=} \{[\rho] \in R \mid \rho(\xi) = -1, \rho(\delta_i) = 1, 1 \le i \le g-1\}. \end{aligned}$$

Note that Q_1^z and Q_2 are representation spaces corresponding to a Heegaard splitting which represents z-surgery on K. Hence

$$\bar{\lambda}(N_z) = \langle Q_1^z, Q_2 \rangle.$$

Assume that Q_2 has been isotoped into general position with respect to $Q_1^{\pm z}$ for all primitive $z \in H_1(\partial U; \mathbf{Z})$, $\langle z, l \rangle \neq 0$.

Let h be a left-handed Dehn twist along β. As usual, let h also denote the induced maps on homology groups, homotopy groups and representation spaces. One can easily verify that $h(a) = a + b$, and hence that

$$h(Q_1^{a+b}) = Q_1^a.$$

Following the pattern of previous lemmas, we will, to the extent possible, represent $h(Q_1^{a+b}) - Q_1^{a+b}$ as a "difference cycle". This will be more delicate than in previous cases, since $h|_{T_1} \neq \mathrm{id}$.

Let F' be F cut along β. Then $\pi_1(F)$ is generated by $\pi_1(F')$ and α. If base points are chosen properly, the action of h on $R^{\natural}$ is given by

$$\begin{aligned} h(\rho)x &= \rho(x), & x \in \pi_1(F') \\ h(\rho)\alpha &= \rho(\beta)\rho(\alpha). \end{aligned}$$

Let $f : SU(2) \times [0,1) \to [0,1]$ be a smooth, Ad-invariant function such that $f(X,t) = 1$ if $\mathrm{tr}(X) \geq -1 - t$, $f(-1,t) = 0$ for all t, and $f(\cdot, t)$ is monotonic (on $SU(2)/\mathrm{Ad}$) for all t. For $t \in [0,1)$, define $h_t^{\natural} : R^{\natural} \to R^{\natural}$ by

$$\begin{aligned} h_t^{\natural}(\rho)x &= \rho(x), & x \in \pi_1(F') \\ h_t^{\natural}(\rho)\alpha &= \rho(\beta)^{f(\rho(\beta),t)t}\rho(\alpha). \end{aligned}$$

$\{h_t^{\natural}\}$ descends to an isotopy $\{h_t\}$ of R.

Arguing as in the proof of (3.2), we see that as $t \to 1$, $h_t(Q_1^{a+b}) \to Q_1^a \cup Q_1^{-b}$. For notational convenience, let $h_1(Q_1^{a+b}) \overset{\text{def}}{=} Q_1^a \cup Q_1^{-b}$. Note also that $h_0 = \mathrm{id}$ and that as $t \to 1$, $h_t \to h$ on $\{[\rho] \in R \,|\, \rho(\beta) \neq -1\}$. Hence Q_1^{-b} will play the role of the difference cycle. To make this precise will require closer examination of the singularities of R.

First we show that $\{h_t\}$ is a not-so-special isotopy (cf (2.16)). The conditions on a, b and l guarantee that $h_t(Q_1^{a+b})$ is transverse to Q_2 at P. By Theorem 4.7 of [G3], h_t is a symplectic map of R to itself, and so h_t acts symplectically on ν. All that remains to be shown is that $h_t(\widetilde{T_1})$ is transverse to $\widetilde{T_2}$.

Pick curves κ_i, $1 \leq i \leq g - 1$, so that $\{\delta_i, \kappa_i, \alpha, \beta\}$ is a standard symplectic set of curves on F. Let

$$X \overset{\text{def}}{=} \{\rho \in \widetilde{S} \,|\, \rho(\delta_i) = 1, 1 \leq i \leq g - 1\} \cong (S_0^1)^{g+1}$$

$$V \overset{\text{def}}{=} \{\rho \in \widetilde{S} \,|\, \rho(\delta_i) = \rho(\kappa_i) = 1, 1 \leq i \leq g - 1\} \cong (S_0^1)^2.$$

Let $\pi : X \to V$ be the projection (with respect to the coordinates induced from $\{\delta_i, \kappa_i, \alpha, \beta\}$). Use $A \overset{\text{def}}{=} \rho(\alpha)$ and $B \overset{\text{def}}{=} \rho(\beta)^{-1}$ as coordinates on V. Then

$$\pi(\widetilde{T_1}^a) = \widetilde{T_1}^a \cap V = \{(A, B) \in V \mid A = 1\}$$

$$\pi(\widetilde{T_1}^{\pm b}) = \widetilde{T_1}^{\pm b} \cap V = \{(A, B) \in V \mid B = \pm 1\}$$

$$\pi(\widetilde{T_1}^{a+b}) = \widetilde{T_1}^{a+b} \cap V = \{(A, B) \in V \mid A - B = 1\}$$

$$(3.37) \qquad \pi(\widetilde{T_2} \cap X) = \{(A, B) \in V \mid uA + vB = 1\},$$

for some integers $u, v > 0$. (Note that we are writing group composition in S_0^1 additively.) $\pi : \widetilde{T_2} \cap X \to \pi(\widetilde{T_2} \cap X)$ is a covering. Let n be its degree. (If we ignore the fact that we are dealing with tori rather than vector spaces, then $\pi(\widetilde{T_2} \cap X)$ is the symplectic reduction of the lagrangian torus $\widetilde{T_2}$ with respect to the coisotropic torus X (see [We]).) Note that

$$(3.38) \qquad |H_1(N_a; \mathbf{Z})| = |\widetilde{T_1}^a \cap \widetilde{T_2}| = nv$$

$$(3.39) \qquad |H_1(N_b; \mathbf{Z})| = |\widetilde{T_1}^b \cap \widetilde{T_2}| = nu$$

$h_t(\widetilde{T_1}^{a+b}) \subset X$ for all t, and $\pi(h_t(\widetilde{T_1}^{a+b})) = h_t(\widetilde{T_1}^{a+b}) \cap V$. Thus $h_t(\widetilde{T_1}^{a+b})$ is transverse to $\widetilde{T_2}$ if and only if $h_t(\widetilde{T_1}^{a+b}) \cap V$ is transverse to $\pi(\widetilde{T_2} \cap X)$. This follows from the monotonicity of the function f used in defining h_t. This completes the proof that h_t is a not-so-special isotopy.

Let $\mathcal{W}_0$ be a wiring of Q_1^{a+b} and Q_2. Extend this to a continuous, 1-parameter family $\mathcal{W}_t \overset{\text{def}}{=} h_t(\mathcal{W}_0)$ of wirings of $h_t(Q_1^{a+b})$ and Q_2, $0 \le t < 1$. By (2.16),

$$(3.40) \qquad A_1(h_t(Q_1^{a+b}), Q_2, \mathcal{W}_t) = A_1(Q_1^{a+b}, Q_2, \mathcal{W}_0)$$

for all $0 \le t < 1$. Our next goal is to make sense of (3.40) when $t = 1$.

Recall that as $t \to 1$, $h_t(Q_1^{a+b})$ tends to $h_1(Q_1^{a+b}) = Q_1^a \cup Q_1^{-b}$. The paths and surfaces of $\mathcal{W}_t$ can be chosen so that they have limits as $t \to 1$ and so that the $\widetilde{T_1}$ paths consist of an initial segment which lies in $\pi^{-1}(1)$ and a final segment which projects 1 to 1 into V and on which $\rho(\delta_i)$ is constant, $1 \le i \le g - 1$ (see Figure 3.15). Define $\mathcal{W}_1$ to be the limit of $\mathcal{W}_t$ as $t \to 1$. Then $\mathcal{W}_1$ restricts to wirings $\mathcal{W}_1^a$ of Q_1^a and Q_2 and $\mathcal{W}_1^{-b}$ of Q_1^{-b} and Q_2. (If $p \in \widetilde{T_1}^a \cap \widetilde{T_1}^{-b} \cap \widetilde{T_2}$, then there are two points of $h_t(\widetilde{T_1}^{a+b}) \cap \widetilde{T_2}$ which tend to p, and correspondingly two configurations of arcs and surface in $\mathcal{W}_1$. Arbitrarily assign one of these to $\mathcal{W}_1^a$ and the other to $\mathcal{W}_1^{-b}$.)

We will now define $A_1(Q_1^{-b}, Q_2, W_1^{-b})$ in such a way that

$$(3.41) \quad A_1(Q_1^a, Q_2, W_1^a) + A_1(Q_1^{-b}, Q_2, W_1^{-b}) = A_1(Q_1^{a+b}, Q_2, W_0).$$

Let $p \in T_1^{-b} \cap T_2$ and let $p', p'' \in \widetilde{T_1}^{-b} \cap \widetilde{T_2}$ be the inverse images of p in $\widetilde{S}$. (If $p \in P$, then $p' = p''$.) Let $\alpha_1' \in W_1^{-b}$ be the arc from 1 to p' in $\widetilde{T_1}^a \cup \widetilde{T_1}^{-b}$. α_1' consists of an initial segment in $\widetilde{T_1}^a$ and a final segment in $\widetilde{T_1}^{-b}$. (If $p' \in \widetilde{T_1}^a \cap \widetilde{T_1}^{-b}$, then the final segment consists only of p'.) The normal bundles of $\widetilde{T_1}^a$ and $\widetilde{T_1}^{-b}$ determine sections of $\det^1(\nu)$ over these two segments, but the sections do not coincide in the middle. Patch them together in the middle so that the relative homotopy class of the resulting section agrees with the relative homotopy class of the limit as $t \to 1$ of the corresponding sections over the appropriate arcs in W_t. Define a section of $\det^1(\nu)$ over α_1'' similarly. This done, we can now proceed as in (2.A) to define the trivializations $\Phi_\pm$. Now define $I_1(p, W_1^{-b})$ by (2.3) and $A_1(Q_1^{-b}, Q_2, W_1^{-b})$ by (2.4).

We now establish (3.41). Since Q_2 is in general position with respect to $h_1(Q_1^{a+b})$, Q_2 is in general position with respect to $h_t(Q_1^{a+b})$ for all t sufficiently close to 1. Hence

$$(3.42) \quad \sum_{p \in h_t(Q_1^{a+b})-\cap Q_2^-} \operatorname{sign}(p) = \sum_{p \in (Q_1^a)-\cap Q_2^-} \operatorname{sign}(p) + \sum_{p \in (Q_1^{-b})-\cap Q_2^-} \operatorname{sign}(p)$$

for t near 1. If it can be shown that

$$(3.43) \quad \sum_{p \in h_t(T_1^{a+b})\cap T_2} I_1(p, W_t) = \sum_{p \in T_1^a \cap T_2} I_1(p, W_1^a) + \sum_{p \in T_1^{-b} \cap T_2} I_1(p, W_1^{-b})$$

for t near 1, then (3.41) will follow from (3.42), (3.43), (3.40) and (2.4).

Let $p' \in h_t(\widetilde{T_1}^{a+b}) \cap \widetilde{T_2} \subset \widetilde{S}$ and let $p \in S$ be the corresponding point downstairs. Define

$$\widetilde{I}_1(p', W_t) = \begin{cases} \frac{1}{2} I_1(p, W_t), & p \notin P \\ I_1(p, W_t), & p \in P. \end{cases}$$

Make similar definitions for $p' \in \widetilde{T_1}^a \cap \widetilde{T_2}$ and $p' \in \widetilde{T_1}^{-b} \cap \widetilde{T_2}$. Then (3.43) is equivalent to

$$(3.44) \quad \sum_{p' \in h_t(\widetilde{T_1}^{a+b})\cap \widetilde{T_2}} \widetilde{I}_1(p', W_t) = \sum_{p' \in \widetilde{T_1}^a \cap \widetilde{T_2}} \widetilde{I}_1(p', W_1^a) + \sum_{p' \in \widetilde{T_1}^{-b} \cap \widetilde{T_2}} \widetilde{I}_1(p', W_1^{-b})$$

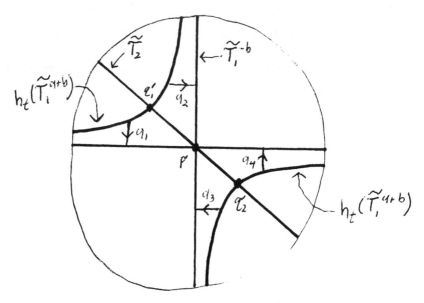

Figure 3.13: A neighborhood of $p' \in \widetilde{T_1}^{-a} \cap \widetilde{T_1}^{-b} \cap \widetilde{T_2}$, projected onto V.

Let $p' \in (\widetilde{T_1}^{-b} \cap \widetilde{T_2}) \setminus \widetilde{T_1}^{-a}$ and let q' be the corresponding point in $h_t(\widetilde{T_1}^{a+b}) \cap \widetilde{T_2}$. It is clear from the definition of $I_1(p, \mathcal{W}_1^{-b})$ that

$$\widetilde{I}_1(p', \mathcal{W}_1^{-b}) = \widetilde{I}_1(q', \mathcal{W}_t)$$

for t near 1. If $p' \in (\widetilde{T_1}^{a} \cap \widetilde{T_2}) \setminus \widetilde{T_1}^{-b}$, the corresponding statement is even clearer. Now let $p' \in \widetilde{T_1}^{a} \cap \widetilde{T_1}^{-b} \cap \widetilde{T_2}$. In this case there are two points of $h_t(\widetilde{T_1}^{a+b}) \cap \widetilde{T_2}$, q_1' and q_2', which tend to p' as $t \to 1$ (see Figure 3.13). (3.44) will be established if we can show that

$$(3.45) \qquad \widetilde{I}_1(p', \mathcal{W}_1^{a}) + \widetilde{I}_1(p', \mathcal{W}_1^{-b}) = \widetilde{I}_1(q_1', \mathcal{W}_t) + \widetilde{I}_1(q_2', \mathcal{W}_t)$$

for t near 1.

Let $U \subset \widetilde{S}$ be a neighborhood of p' containing q_1' and q_2'. Let

$$\Psi : \det{}^1(\nu)|_U \to S^1$$

be a trivialization. Let the arcs a_1, a_2, a_3 and $a_4 \subset U$ be as shown in Figure 3.13. $h_t(\eta_1^{a+b})$, for various t near 1, determines a section of $\det^1(\nu)$ over a_i, $1 \le i \le 4$. Thus all of the subspaces pictured in Figure 3.13 have

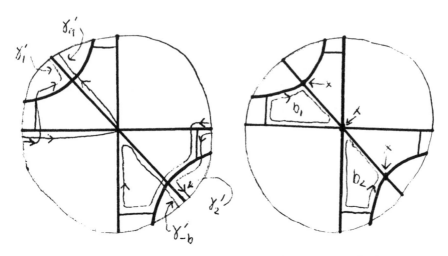

Figure 3.14: Non-coinciding parts of the homotoped γ's.

sections of $\det^1(\nu)$ defined over them. These sections do not agree at p', q_1'
and q_2', but they can be patched together using the paths $P_\pm$ in $\mathcal{L}(\nu)$ (see
(2.A)).

Without loss of generality, the limit of the part of $\mathcal{W}_t$ associated to q_1'
is in $\mathcal{W}_1^a$ and the limit of the part of $\mathcal{W}_t$ associated to q_2' is in $\mathcal{W}_1^{-b}$. Let γ_1',
γ_2', γ_a' and γ_{-b}' be the loops in $\mathcal{W}_t$, $\mathcal{W}_t$, $\mathcal{W}_1^a$ and $\mathcal{W}_1^{-b}$, respectively, through
q_1', q_2', p' and p', respectively. These loops can be homotoped, together with
their sections of $\det^1(\nu)$, so that γ_1' (and its section) coincides with γ_a' and
γ_2' coincides with γ_{-b}', except in U, where they differ as shown in Figure
3.14. We may assume that the sections over the (homotoped) γ's in Figure
3.14 agree with the sections defined in the previous paragraph.

Let b_1 [b_2] be the loop in U representing $\gamma_1' - \gamma_a'$ [$\gamma_2' - \gamma_{-b}'$], as shown in
Figure 3.14. Using the paths $P_\pm$ (over the points marked 'x' in the figure),
we get sections of $\det^1(\nu)$ over b_1 and b_2. Let $e_1^\pm$ and $e_2^\pm$ denote the degree
of these sections with respect to the trivialization Ψ. Let $\Phi_{1\pm}'$, $\Phi_{2\pm}'$, $\Phi_{a\pm}'$
and $\Phi_{-b\pm}'$ be the sections of $\det^1(\nu)$ over γ_1', γ_2', γ_a' and γ_{-b}'. Then we have

$$
\begin{aligned}
\Phi_{1\pm}' - \Phi_{a\pm}' &= e_1^\pm \\
\Phi_{2\pm}' - \Phi_{-b\pm}' &= e_2^\pm.
\end{aligned}
$$

(The trivialization Ψ of $\det^1(\nu)$ over U is used to compare the sections.)
Therefore (3.45) follows from

(3.46) $$e_1^+ + e_1^- + e_2^+ + e_2^- = 0.$$

This will be proved by a symmetry argument.

Note that $e_i^{\pm}$ depends only on the intersections with U of $h_t(\widetilde{T_1}^{a+b})$, $\widetilde{T_1}^a$, $\widetilde{T_1}^{-b}$ and $\widetilde{T_2}$, and their normal bundles. Note also that it does not change $e_i^{\pm}$ to vary $\widetilde{T_2}$ and its normal bundle, so long as they are kept in general position with respect to $h_t(\widetilde{T_1}^{a+b})$, $\widetilde{T_1}^a$, $\widetilde{T_1}^{-b}$ and their normal bundles.

Let $H^1 \overset{\text{def}}{=} H_1(F; \mathbf{R})$. It follows from (1.13) that $\nu|_U$ can be identified with a subset of $(H^1 \otimes \mathbf{h}) \times (H^1 \otimes \mathbf{h}^{\perp})$. (If $p' \notin P$, then p' will not correspond to $0 \in H^1 \otimes \mathbf{h}$.) Let $\bar{\alpha}$, $\bar{\beta}$, $\bar{\delta_i}$, $\bar{\kappa_i}$ be the coordinates on H^1 induced by the basis α, β, δ_i, κ_i of H^1. So, for example, $\widetilde{T_1}^{-b} \cap U$ corresponds to $\{(\bar{\alpha}, \bar{\beta}, \bar{\delta_i}, \bar{\kappa_i}) \in H^1 \,|\, \bar{\alpha} = 0, \bar{\delta_i} = 0, 1 \le i \le g - 1\} \otimes \mathbf{h}$.

Consider the involution σ of H^1 given by

$$
\begin{aligned}
\bar{\alpha} &\mapsto \bar{\beta} \\
\bar{\beta} &\mapsto \bar{\alpha} \\
\bar{\delta_i} &\mapsto \bar{\delta_i} \\
\bar{\kappa_i} &\mapsto -\bar{\kappa_i}
\end{aligned}
$$

σ preserves the cup product pairing on H^1 up to sign, and so induces an involution (also denoted σ) of $\nu|_U$. σ interchanges $\widetilde{T_1}^a \cap U$ and $\widetilde{T_1}^{-b} \cap U$ (and their normal bundles). $\widetilde{T_2} \cap U$ and its normal bundle can be isotoped (keeping things in general position) so that they are preserved by σ. h_t can be chosen so that $h_t(\widetilde{T_1}^{a+b}) \cap U$ and its normal bundle are preserved by σ. σ interchanges the loops b_1 and b_2. Since $\sigma^* \omega = -\omega$, σ interchanges the paths $\det^1(P_+)$ and $\det^1(P_-)$ (in the fibers of $\det^1(\nu)$ over p', q_1' and q_2'), and also reverses the orientation of $\det^1(\nu)|_U$. Therefore

$$
e_1^+ + e_1^- = -(e_2^+ + e_2^-),
$$

This completes the proof of (3.41).

Our next goal is to show that

$$
\begin{aligned}
(3.47) \quad A_2(Q_1^{a+b}, Q_2, W_0) = {}& A_2(Q_1^a, Q_2, W_1^a) + A_2(Q_1^{-b}, Q_2, W_1^{-b}) \\
& + \frac{1}{6}(|H_1(N_a; \mathbf{Z})| - |H_1(N_b; \mathbf{Z})|) + \frac{1}{2}|H_1(N_b; \mathbf{Z})|.
\end{aligned}
$$

This is just a matter of computing the change in $\int_E \omega$ for various surfaces E in W_0 and W_1. We will do this for the case where u is odd and v is even, the other three cases being similar.

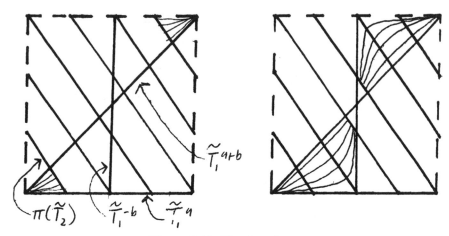

Figure 3.15: Tracks of arcs.

Note that if E is any surface in X, then

$$\int_E \omega = \int_{\pi(E)} \omega.$$

If E_0 is a surface in $\mathcal{W}_0$ and E_1 is the corresponding surface in $\mathcal{W}_1$, then we may assume that E_0 and E_1 differ by the track of the arcs α'_1 and α''_1 under the isotopy $\{h_t\}$. So we must add up the ω-areas of these tracks after projecting them onto V. Two typical examples are shown in Figure 3.15.

Identify V with $I \times I / ((0, x) \sim (1, x), (x, 0) \sim (x, 1))$. Then

$$\pi(\widetilde{T_2} \cap X) = \{(A, B) \in I \times I \mid uA + vB = k, k \in \mathbf{Z}, 0 \le k < u + v\}.$$

Thus to each point of $\widetilde{T_1}^{a+b} \cap \widetilde{T_2}$ there is associated an integer k, $0 \le k < u + v$. Since $\pi : \widetilde{T_2} \cap X \to \pi(\widetilde{T_2} \cap X)$ is an n-fold covering, each k is associated to n points of $\widetilde{T_1}^{a+b} \cap \widetilde{T_2}$.

Let $p' \in \widetilde{T_1}^{a+b} \cap \widetilde{T_2}$ and let k be the associated integer. Assume $0 \le k \le (u + v - 1)/2$. Let α'_1 be the path of $\mathcal{W}_0$ from 1 to p'. Let ΔE be the track of $h_t(\alpha'_1)$. Keeping in mind that $\int_V \omega = -8$ (see (1.15)), it is not hard to see that

$$\int_{\pi(\Delta E)} \omega = \begin{cases} \frac{4k^2}{v(u+v)}, & 0 \le k \le v/2 \\ 1 - \frac{4}{u(u+v)}(\frac{u+v}{2} - k)^2, & v/2 \le k \le (u + v - 1)/2. \end{cases}$$

Therefore

$$A_2(Q_1^{a+b}, Q_2, \mathcal{W}_0) - A_2(Q_1^a, Q_2, \mathcal{W}_1^a) - A_2(Q_1^{-b}, Q_2, \mathcal{W}_0^{-b})$$

$$= n\left(\sum_{k=0}^{v/2}\frac{4k^2}{v(u+v)} + \sum_{k=v/2+1}^{(u+v-1)/2}(1 - \frac{4}{u(u+v)}(\frac{u+v}{2} - k)^2)\right)$$

$$= n\left(\frac{4}{v(u+v)}\frac{v(v+2)(v+1)}{24} + \frac{u-1}{2}\right.$$

$$\left. - \frac{4}{u(u+v)}\left(\frac{(u^2-1)u}{24} - \frac{(u^2-1)}{8} + \frac{(u-1)}{8}\right)\right)$$

$$= n\left(\frac{1}{6}(v-u) + \frac{1}{2}u\right)$$

$$= \frac{1}{6}(|H_1(N_a;\mathbf{Z})| - |H_1(N_b;\mathbf{Z})|) + \frac{1}{2}|H_1(N_b;\mathbf{Z})|.$$

This completes the proof of (3.47).

Define

$$\langle Q_1^{-b}, Q_2\rangle \stackrel{\text{def}}{=} A_1(Q_1^{-b}, Q_2, W_1^{-b}) + A_2(Q_1^{-b}, Q_2, W_1^{-b}).$$

It follows from (3.41) and (3.47) that

$$r(a,b) = \bar\lambda(N_{a+b}) - \bar\lambda(N_a) - \bar\lambda(N_b) - \frac{1}{6}(|H_1(N_a;\mathbf{Z})| - |H_1(N_b;\mathbf{Z})|)$$

$$= (A_1(Q_1^{a+b}, Q_2, W_0) + A_2(Q_1^{a+b}, Q_2, W_0) - A_1(Q_1^a, Q_2, W_1^a)$$

$$- A_2(Q_1^a, Q_2, W_1^a) - \langle Q_1^b, Q_2\rangle) - \frac{1}{6}(|H_1(N_a;\mathbf{Z})| - |H_1(N_b;\mathbf{Z})|)$$

$$(3.48) \quad = (\langle Q_1^{-b}, Q_2\rangle - \langle Q_1^b, Q_2\rangle) + \frac{1}{2}|H_1(N_b;\mathbf{Z})|.$$

Now, finally, we show that $r(a+b,b) = r(a,b)$. The transformation $a \mapsto a+b$, $b \mapsto b$ is effected by the Dehn twist h. Applying (3.48) and keeping in mind that $h(Q_1^{\pm b}) = Q_1^{\pm b}$, we see that

$$r(a+b,b) - r(a,b) = \langle h(Q_1^{-b}), Q_2\rangle - \langle Q_1^{-b}, Q_2\rangle$$

$$= A_1(Q_1^{-b}, Q_2, V_1^{-b}) - A_1(Q_1^{-b}, Q_2, W_1^{-b})$$

$$+ A_2(Q_1^{-b}, Q_2, V_1^{-b}) - A_2(Q_1^{-b}, Q_2, W_1^{-b}),$$

where V_1^{-b} is induced from some wiring V_0 of $Q_1^{a+2b} = h(Q_1^{a+b})$ and Q_2, as above.

Let $p' \in \widetilde{T_1}^{-b} \cap \widetilde{T_2}$. Let α_1' and β_1' be the paths of W_1^{-b} and V_1^{-b} connecting p' to 1 through $\widetilde{T_1}^{-b}$. We may assume that α_1' and β_1' differ

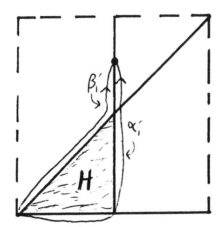

Figure 3.16: $\partial^{-1}(\alpha'_1 - \beta'_1)$.

by the boundary of a triangle $H \subset X$ parallel to V as shown in Figure 3.16. Recall that τ is the covering involution of $\widetilde{S}$. We may assume that $E^\dagger = E \cup H \cup \tau(H)$, where E and $E^\dagger$ are the surfaces in $\mathcal{W}_1^{-b}$ and $\mathcal{V}_1^{-b}$ associated to p' (and $p'' = \tau(p')$).

Using the path $P_\pm$ to patch things up at the trivial representation, we get a trivialization (induced by $\det^1(\eta_1)|_{\alpha'_1}$ and $\det^1(h(\eta_1))|_{\beta'_1}$) of $\det^1(\nu)|_{\partial H}$. Let $e_\pm \in \mathbf{Z}$ be difference between this trivialization and the one induced from $\det^1(\nu)|_H$. It is easy to see that

$$\tilde{I}_1(p', \mathcal{V}_1^{-b}) - \tilde{I}_1(p', \mathcal{W}_1^{-b}) = \frac{e_+ + e_-}{4}.$$

Since $\widetilde{T_1}^{-b} \cap \widetilde{T_2}$ consists of $|H_1(N_b; \mathbf{Z})|$ points,

$$A_1(Q_1^{-b}, Q_2, \mathcal{V}_1^{-b}) - A_1(Q_1^{-b}, Q_2, \mathcal{W}_1^{-b}) = \frac{e_+ + e_-}{4}|H_1(N_b; \mathbf{Z})|.$$

Similarly,

$$\tilde{I}_2(p', \mathcal{V}_1^{-b}) - \tilde{I}_2(p', \mathcal{W}_1^{-b}) = \frac{1}{2}\int_H \omega = \frac{1}{2}$$

and

$$A_2(Q_1^{-b}, Q_2, \mathcal{V}_1^{-b}) - A_2(Q_1^{-b}, Q_2, \mathcal{W}_1^{-b}) = \frac{1}{2}|H_1(N_b; \mathbf{Z})|.$$

Therefore

(3.49) $$r(a + b, b) - r(a, b) = |H_1(N_b; \mathbf{Z})|(\frac{e_+ + e_-}{4} + \frac{1}{2}).$$

So what must be shown is that

$$x \stackrel{\mathrm{def}}{=} \frac{e_+ + e_-}{4} + \frac{1}{2} = 0.$$

Note that x is independent of Q_2, and hence of K, a and b. It therefore suffices to show that x is zero for some particular knot K and homology classes a, b.

Let K be the unknot in S^3. Let $N = S^3 \setminus \mathrm{nbd}(K)$. Let $b \in H_1(\partial(N); \mathbf{Z})$ be a meridian and let $l \in H_1(\partial(N); \mathbf{Z})$ be a longitude, oriented so that $\langle l, b \rangle = 1$. Then

$$N_b \cong S^3$$

and, for $p \in \mathbf{Z}$,

$$N_{pb+l} \cong -L_{p,1},$$

where $L_{p,1}$ denotes the $p, 1$ lens space, i.e. $p/1$ surgery on K.
    Let

$$r_0 \stackrel{\mathrm{def}}{=} r(b + l, b).$$

Then, by (3.49),

$$r(pb + l, b) = r_0 + x(p - 1)$$

for $p \geq 2$. Hence

$$
\begin{aligned}
\bar{\lambda}(L_{p,1}) &= -\bar{\lambda}(N_{pb+l}) \\
&= -\bar{\lambda}(N_{(p-1)b+l}) - \bar{\lambda}(N_b) - \frac{1}{6}(|H_1(N_{(p-1)b+l}; \mathbf{Z})| - |H_1(N_b; \mathbf{Z})|) \\
&\qquad - r((p-1)b + l, b) \\
&= \bar{\lambda}(L_{p-1,1}) - \frac{1}{6}(p-2) - r_0 - x(p-2) \\
&= \bar{\lambda}(L_{p-1,1}) + (-\frac{1}{6} - x)p + (\frac{1}{3} - r_0 + 2x).
\end{aligned}
$$

Therefore, using the fact that $\bar{\lambda}(L_{1,1}) = \bar{\lambda}(S^3) = 0$,

$$\bar{\lambda}(L_{p,1}) = (-\frac{1}{12} - \frac{x}{2})p^2 + (-r_0 + \frac{1}{4} + \frac{3x}{2})p + (r_0 - \frac{1}{6} - x).$$

But, by (3.18) and (3.21),

$$
\begin{aligned}
\bar{\lambda}(L_{2,1}) &= 0 \\
\bar{\lambda}(L_{5,1}) &= \mp 1.
\end{aligned}
$$

Solving for r_0 and x, we get

$$
\begin{aligned}
r_0 &= 0 \\
x &= \frac{\pm 1 - 1}{6}.
\end{aligned}
$$

Since $4x \in \mathbf{Z}$, we must have $\bar{\lambda}(L_{5,1}) = -1$ and $x = 0$. This completes the proof that $r(a+b, b) = r(a, b)$. The proof that $r(a, a+b) = r(a, b)$ is similar. The proof of (3.36) is complete.   □

As a byproduct of the previous proof we proved that $\bar{\lambda}(L_{5,1}) = -1$. The proof of (3.21) shows that this is equivalent to

(3.50) **Lemma.** *Let K be the right-handed trefoil knot in S^3. Then $\lambda'(K) = 2$.*   □

(3.51) **Lemma.** *Let $K \subset M$ be an unknot in a* **Q***HS. Then $r(K, a, b) = 0$ for all appropriate $a, b \in H_1(\partial(\mathrm{nbd}(K)); \mathbf{Z})$ (see (3.36)).*

Proof: As usual, let $N \stackrel{\text{def}}{=} M \setminus \mathrm{nbd}(K)$. Let $m, l \in H_1(\partial N; \mathbf{Z})$ be a standard meridian-longitude basis of $H_1(\partial N; \mathbf{Z})$. It suffices to show that $r(K, m+il, m+(i+1)l) = 0$ for all $i \in \mathbf{Z}$. Since $N_{pm+ql} \cong N_{pm+(ip+q)l}$ for all $i \in \mathbf{Z}$ and relatively prime $p, q \in \mathbf{Z}$, it suffices to show that $r(K, m, m+l) = 0$. But $N_m \cong N_{m+l} \cong M$, while $N_{2m+l} \cong M \# \mathbf{R}P^2$. Therefore (3.51) is implied by

(3.52) $\bar{\lambda}(M \# \mathbf{R}P^3) = 2\bar{\lambda}(M)$ *for any* **Q***HS M.*

Let (W_{1a}, W_{2a}, F_a) be a Heegaard splitting for M and let (W_{1b}, W_{2b}, F_b) be the genus one Heegaard splitting of $\mathbf{R}P^3$. Let (W_1, W_2, F) be the connected sum of these two Heegaard splittings. Appropriate appropriate notation from the proof of (3.22) (e.g. R_a, R_b, Z, $q : Z \to R_a \times R_b$, etc.).

As in the proof of (3.22), we may assume that

$$Q_j = q^{-1}(Q_{aj} \times Q_{bj})$$

holds with Q_{1a} transverse to Q_{2a} in R_a. (Note that Q_{1b} is automatically transverse to Q_{2b} in R_b; no isotoping is required.) Therefore

$$Q_1 \cap Q_2 = \bigcup_{(p_a, p_b)} q^{-1}(p_a, p_b),$$

where (p_a, p_b) runs through $(Q_{1a} \cap Q_{2a}) \times (Q_{1b} \cap Q_{2b})$. $(Q_{1b} \cap Q_{2b})$ consists of two points: the trivial representation 1_b and a non-trivial representation $\sigma \in P_b$. It follows that, for each (p_a, p_b), $q^{-1}(p_a, p_b)$ consists of a single point, and that Q_1 and Q_2 are transverse. Note that $q^{-1}(p_a, p_b)$ lies in R^-,

S^- or P exactly when p_a lies in R_a^-, S_a^- or P_a, respectively. For notational simplicity, we will henceforth denote $q^{-1}(p_a, p_b)$ by (p_a, p_b).

We have

$$\bar{\lambda}(M \# \mathbf{R}P^3) = \sum_{p \in Q_{1a}^- \cap Q_{2a}^-} (\text{sign}\,(p, 1_b) + \text{sign}\,(p, \sigma))$$

$$+ \sum_{p \in T_{1a} \cap T_{2a}} (I(p, 1_b) + I(p, \sigma)).$$

The proof will be complete if it can be shown that

$$(3.53) \qquad \bar{\lambda}(M) = \sum_{p \in Q_{1a}^- \cap Q_{2a}^-} \text{sign}\,(p, 1_b) + \sum_{p \in T_{1a} \cap T_{2a}} I(p, 1_b)$$

and that

$$(3.54) \qquad\qquad\qquad \text{sign}\,(p, 1_b) = \text{sign}\,(p, \sigma)$$
$$(3.55) \qquad\qquad\qquad\quad I(r, 1_b) = I(r, \sigma)$$

for all $p \in Q_{1a}^- \cap Q_{2a}^-$ and $r \in T_{1a} \cap T_{2a}$.

The proof of (3.54) is similar to the proof of the corresponding fact in the proof of (3.22), and so is omitted.

$R_a \times \{1_b\}$ and $R_a \times \{\sigma\}$ can be regarded as subspaces of R. Arguing as in the proof (in (2.B)) of the stabilization invariance of λ, we see that $R_a \times \{p\}$ has a trivial normal bundle in R ($p = 1_b$ or σ). Similarly, $Q_{ja} \times \{p\}$ has a trivial normal bundle in Q_j. This implies (3.53). It also implies that a neighborhood of the triple $(S_a, T_{1a}, T_{2a}) \times \{1_b\}$ in (R, Q_1, Q_2) is diffeomorphic to a neighborhood of the triple $(S_a, T_{1a}, T_{2a}) \times \{\sigma\}$ in (R, Q_1, Q_2).

For $r \in T_{1a} \cap T_{2a}$, let $I((1_a, \sigma), (r, \sigma))$ be the same as $I(r, \sigma)$, but with $(1_a, \sigma)$ playing the role of $1 = (1_a, 1_b)$ (i.e. the paths of the wiring connect (r, σ) to $(1_a, \sigma)$). Clearly

$$(3.56) \qquad\qquad\qquad I(r, 1_b) = I((1_a, \sigma), (r, \sigma)).$$

On the other hand, it is easy to see that

$$(3.57) \qquad\qquad I(r, \sigma) = x I(1_a, \sigma) + I((1_a, \sigma), (r, \sigma)),$$

where $x = 1$ if $r \in P_a$ and $x = 2$ if $r \notin P_a$. Since $\{1_a\} \times (R_b, Q_{1b}, Q_{2b})$ has trivial normal bundle in (R, Q_1, Q_2),

$$\lambda(\mathbf{R}P^3) = \frac{I(1_a, 1_b) + I(1_a, \sigma)}{2}.$$

But $\lambda(\mathbf{R}P^3) = 0$ (by (3.1)) and $I(1_a, 1_b) = 0$. Hence

(3.58) $$I(1_a, \sigma) = 0.$$

(3.55) now follows from (3.56), (3.57) and (3.58). □

(3.59) **Lemma.** *Let K and L be knots in a $\mathbf{Q}HS$ M such that L is null-homologous and does not link K (i.e. L bounds a Seifert surface disjoint from K). Let M_n denote $1/n$ Dehn surgery along L. Let $r(K, M_n; \; \cdot \; , \; \cdot \;)$ denote the function introduced in (3.36) (defined on a certain subset of $H_1(\partial(\mathrm{nbd}(K)); \mathbf{Z}) \times H_1(\partial(\mathrm{nbd}(K)); \mathbf{Z}))$, with K thought of as a knot in M_n. Then $r(K, M_n; \; \cdot \; , \; \cdot \;)$ is independent of n.*

Proof: Retain all notation from the proof of (3.36).

Note that since $\mathrm{lk}\,(K, L) = 0$, surgery on L does not affect the longitude of K, and so the domain of $r(K, M_n; \; \cdot \; , \; \cdot \;)$ is independent of n. Let $(a, b) \in H_1(\partial(\mathrm{nbd}(K)); \mathbf{Z}) \times H_1(\partial(\mathrm{nbd}(K)); \mathbf{Z})$ be in the domain (i.e. $\langle a, b \rangle = 1$ and the longitude of K is not a nonnegative linear combination of a and b).

Choose the Heegaard splitting for M so that, in addition to satisfying the conditions specified in the proof of (3.36) (i.e. W_1 contains $\mathrm{nbd}(K)$ as a boundary connected summand), L coincides with a separating curve $\gamma \subset F$, disjoint from α and β.

Let $h : F \to F$ be a left-handed Dehn twist along γ. As usual, let h also denote the various maps which it induces. Let

$$N \overset{\text{def}}{=} \{[\rho] \in R \,|\, \rho(\gamma) = -1\}.$$

Let $D \subset N$ be the difference cycle representing $h(Q_2) - Q_2$, as constructed in (3.2) (with the roles of Q_1 and Q_2 interchanged).

By (3.48),

$$r(K, M_n; a, b) = \langle Q_1^{-b}, h^n(Q_2) \rangle - \langle Q_1^b, h^n(Q_2) \rangle + \frac{1}{2}|H_1(M_{nb}; \mathbf{Z})|,$$

where M_{nb} denotes $1/n$ surgery on L and b surgery on K. Since

$$|H_1(M_{nb}; \mathbf{Z})| = |H_1(M_{0b}; \mathbf{Z})|$$

for all n,

$$\begin{aligned}
r(K, M_{n+1}; a, b) - r(K, M_n; a, b) &= \langle Q_1^{-b}, h^{n+1}(Q_2) \rangle - \langle Q_1^b, h^{n+1}(Q_2) \rangle \\
&\quad - \langle Q_1^{-b}, h^n(Q_2) \rangle + \langle Q_1^b, h^n(Q_2) \rangle \\
&= \langle Q_1^{-b}, h^n(D) \rangle - \langle Q_1^b, h^n(D) \rangle.
\end{aligned}$$

So the lemma will be proved if it can be shown that

$$\langle Q_1^{-b}, h^n(D) \rangle = \langle Q_1^b, h^n(D) \rangle$$

for all $n \in \mathbf{Z}$.

Let $x_1, y_1, \ldots, x_g, y_g$ be a symplectic basis of $\pi_1(F)$ such that α represents x_1, β represents y_1, and γ represents $\prod_1^k [x_i, y_i]$ for some k. This gives rise to an identification

$$R^{\natural} = \{(X_1, Y_1, \ldots, X_g, Y_g) \in (SU(2))^{2g} \mid \prod_1^g [X_i, Y_i] = 1\}.$$

Define $\tau^{\natural} : R^{\natural} \to R^{\natural}$ by

$$\tau^{\natural}(X_1, Y_1, \ldots, X_g, Y_g) = (X_1, -Y_1, X_2, Y_2, \ldots, X_g, Y_g).$$

$\tau^{\natural}$ descends to a map $\tau : R \to R$.

It is not hard to see that

$$\tau(Q_1^{-b}) = Q_1^b$$

(preserving orientations). By (1.18), τ maps N to itself and

$$\tau_* : H_{3g-3}(N; \mathbf{Z}) \to H_{3g-3}(N; \mathbf{Z})$$

is the identity. In particular, $\tau(h^n(D))$ is homologous to $h^n(D)$ for all n. Therefore

$$
\begin{aligned}
\langle Q_1^{-b}, h^n(D) \rangle &= \langle \tau(Q_1^{-b}), \tau(h^n(D)) \rangle \\
&= \langle Q_1^b, h^n(D) \rangle.
\end{aligned}
$$

$\square$

§ 4

The Dehn Surgery Formula

In this section we state and prove a formula for how λ transforms under a general Dehn surgery (i.e. a Dehn surgery on a **QHS** which yields another **QHS**). Notation from previous sections is *not* retained.

Before stating the Dehn surgery formula, we need some preliminary definitions. Let N be the complement of a knot in a **QHS**. Let $\Delta_N(t^{1/2})$ be the Alexander polynomial of N, normalized so that it is symmetric under $\Delta_N(1) = 1$ and $t^{1/2} \leftrightarrow t^{-1/2}$. Define

$$\Gamma(N) \overset{\text{def}}{=} \frac{d^2}{dt^2}\Delta_N(1).$$

Section B contains a more detailed definition of Γ, as well as the proofs of various properties of Γ used in this section.

Let $a, b, l \in H_1(T^2; \mathbf{Z})$ be such that a and b are primitive (i.e. represented by simple closed curves), $\langle a, l \rangle \neq 0$ and $\langle b, l \rangle \neq 0$. Choose a basis x, y of $H_1(T^2; \mathbf{Z})$ such that $\langle x, y \rangle = 1$ and $l = dy$ for some $d \in \mathbf{Z}$. Define

$$(4.1)\, \tau(a, b; l) \overset{\text{def}}{=} -s(\langle x, a \rangle, \langle y, a \rangle) + s(\langle x, b \rangle, \langle y, b \rangle) + \frac{d^2 - 1}{12} \frac{\langle a, b \rangle}{\langle a, l \rangle \langle b, l \rangle},$$

where $s(q, p)$ denotes the Dedekind sum

$$s(q, p) \overset{\text{def}}{=} (\text{sign}\,(p)) \sum_{k=1}^{|p|} ((k/p))((kq/p))$$

$$((x)) \overset{\text{def}}{=} \begin{cases} 0, & x \in \mathbf{Z} \\ x - [x] - 1/2, & \text{otherwise} \end{cases}$$

Note that $\tau(a, b; l)$ depends only on a, b, l and $\langle \cdot, \cdot \rangle$, not on x or y. Section A contains the proofs of various properties of Γ used in this section.

(4.2) Theorem. *Let $N = M \setminus \mathrm{nbd}(K)$ be the complement of a knot K in a **Q**HS M. Let $l \in H_1(\partial N; \mathbf{Z})$ be the longitude. Let $a, b \in H_1(\partial N; \mathbf{Z})$ be primitive and such that $\langle a, l \rangle \neq 0$, $\langle b, l \rangle \neq 0$. Then*

$$\lambda(N_b) = \lambda(N_a) + \tau(a, b; l) + \frac{\langle a, b \rangle}{\langle a, l \rangle \langle b, l \rangle} \Gamma(N).$$

Note: Casson proved that for K a knot in a **Z**HS, $\lambda'(K) = \Gamma(N)$ (cf (3.7)). The proof of (4.2) is in the same spirit as Casson's, but is much more involved.

Proof: By (5.1), there is some invariant of **Q**HS's, call it λ_c, for which (4.2) holds. ((4.2) could be proved using a result weaker than (5.1), but (5.1) needs to be proved anyway, so why bother proving a redundant weaker result?) Let

$$\delta \overset{\mathrm{def}}{=} \lambda - \lambda_c$$

and

$$\overline{\delta} \overset{\mathrm{def}}{=} |H_1(\cdot ; \mathbf{Z})| \delta = \overline{\lambda} - \overline{\lambda}_c.$$

Then (4.2) is equivalent to

(4.3) *Let $N = M \setminus \mathrm{nbd}(K)$ be the complement of a knot K in a **Q**HS M. Then $\delta(N_a)$ is independent of the primitive homology class $a \in H_1(\partial N; \mathbf{Z})$.*

Let $l = dk$, where $k \in H_1(\partial N; \mathbf{Z})$ is primitive and d is a positive integer. Choose $m \in H_1(\partial N; \mathbf{Z})$ such that $\langle m, k \rangle = 1$. For $(p, q) \in \mathbf{Z}^2$ primitive and $p \neq 0$, define

$$N(p, q) \overset{\mathrm{def}}{=} N_{pm+qk}.$$

By (3.34),

$$\overline{\lambda}(N(1, q)) = \overline{\lambda}(N(1, 0)) + \overline{\lambda}'(N)q$$

for some $\overline{\lambda}'(N) \in \mathbf{Q}$ and all $q \in \mathbf{Z}$. By (4.1) and elementary linear algebra,

$$\overline{\lambda}_c(N(1, q)) = \overline{\lambda}_c(N(1, 0)) + |H_1(N(1, 0); \mathbf{Z})| \left(\frac{\Gamma(N)}{d^2} + \frac{d^2 - 1}{12d^2} \right) q.$$

Therefore

(4.4) $$\overline{\delta}(N(1, q)) = \overline{\delta}_0 + \overline{\delta}' q,$$

where

$$\bar{\delta}_0 \overset{\text{def}}{=} \bar{\lambda}(N(1,0)) - \bar{\lambda}_c(N(1,0))$$

$$\bar{\delta}' \overset{\text{def}}{=} \bar{\lambda}'(N) - |H_1(N(1,0);\mathbf{Z})| \left(\frac{\Gamma(N)}{d^2} + \frac{d^2 - 1}{12d^2} \right).$$

Let $a = p_1 m + q_1 k$, $b = p_2 m + q_2 k \in H_1(\partial N; \mathbf{Z})$ be primitive and such that $\langle a, b \rangle = 1$ and $\langle a, l \rangle, \langle b, l \rangle > 0$ (i.e. $p_1, p_2 > 0$). Let $s = [q_1/p_1] \in \mathbf{Z}$. ($[x]$ denotes the greatest integer $\leq x$.) Then a and b are non-negative linear combinations of $m + sk$ and $m + (s + 1)k$. By (3.36),

$$(4.5) \quad \bar{\lambda}(N(p_1 + p_2, q_1 + q_2)) = \bar{\lambda}(N(p_1, q_1)) + \bar{\lambda}(N(p_2, q_2))$$
$$+ \frac{1}{6}(|H_1(N(p_1, q_1); \mathbf{Z})| - |H_1(N(p_2, q_2); \mathbf{Z})|) + r(s),$$

where

$$r(s) \overset{\text{def}}{=} r(m + sk, m + (s + 1)k).$$

We now prove an analogous statement for $\bar{\lambda}_c$. Let $h \overset{\text{def}}{=} |\text{Tor}(H_1(N; \mathbf{Z}))|$. Then

$$|H_1(N_x; \mathbf{Z})| = |\langle x, l \rangle| h$$

for all primitive $x \in H_1(\partial N; \mathbf{Z})$ not colinear with l. Let a and b be as above. Applying (4.2), we have

$$\bar{\lambda}_c(N_a) = \langle a, l \rangle h \left(\lambda_c(N_{a+b}) + \tau(a + b, a; l) + \frac{\langle a + b, a \rangle}{\langle a + b, l \rangle \langle a, l \rangle} \Gamma(N) \right)$$

$$= \frac{\langle a, l \rangle}{\langle a + b, l \rangle} \bar{\lambda}_c(N_{a+b}) + h \langle a, l \rangle \tau(a + b, a; l) + \frac{h \langle b, a \rangle}{\langle a + b, l \rangle} \Gamma(N).$$

Similarly,

$$\bar{\lambda}_c(N_b) = \frac{\langle b, l \rangle}{\langle a + b, l \rangle} \bar{\lambda}_c(N_{a+b}) + h \langle b, l \rangle \tau(a + b, b; l) + \frac{h \langle a, b \rangle}{\langle a + b, l \rangle} \Gamma(N).$$

Adding these two equations and applying (A.17) yields

$$\bar{\lambda}_c(N_a) + \bar{\lambda}_c(N_b) = \bar{\lambda}_c(N_{a+b}) - \frac{h}{6}(\langle a, l \rangle - \langle b, l \rangle),$$

or

$$(4.6) \quad \bar{\lambda}_c(N(p_1 + p_2, q_1 + q_2)) = \bar{\lambda}_c(N(p_1, q_1)) + \bar{\lambda}_c(N(p_2, q_2))$$
$$+ \frac{1}{6}(|H_1(N(p_1, q_1); \mathbf{Z})| - |H_1(N(p_2, q_2); \mathbf{Z})|).$$

Subtracting (4.6) from (4.5) gives

(4.7) $\overline{\delta}(N(p_1 + p_2, q_1 + q_2)) = \overline{\delta}(N(p_1, q_1)) + \overline{\delta}(N(p_2, q_2)) + r([\frac{q_1}{p_1}]).$

(4.4) and repeated applications of (4.7) now give, for primitive $(p, q) \in \mathbf{Z}^2$ with $p > 0$,

$$\overline{\delta}(N(p, q)) = \overline{\delta}_0 + (\overline{\delta}_0 + r([\frac{q}{p}]))(p - 1) + q\overline{\delta}'.$$

Dividing by $|H_1(N(p,q);\mathbf{Z})| = dhp$ yields

(4.8) $\delta(N(p, q)) = \delta_0 + \dfrac{(p-1)r([\frac{q}{p}])}{dhp} + \dfrac{q\delta'}{p}.$

So what must be shown is that $\delta' = 0$ and that $r(s) = 0$ for all $s \in \mathbf{Z}$. (It follows from (4.2), (5.1) and the fact that $\lambda(S^3) = 0$ that $\delta_0 = 0$ also.)

To proceed further, we need the following special case of (4.3).

(4.9) **Lemma.** *Let $K \subset M$ be a knot in a* **QHS** *such that*

$$[K] \in [\pi_1(M), [\pi_1(M), \pi_1(M)]].$$

Then (4.3) holds for $N = M \setminus \mathrm{nbd}(K)$.   □

The proof will be given later.

Let $M = N(1, 0)$ and let $K \subset M$ be the core of the surgery solid torus (so that $N = M \setminus \mathrm{nbd}(K)$). Let $G = \pi_1(M)$. Since M is a **QHS**, $H_1(M; \mathbf{Z}) = G/[G, G]$ and $[G, G]/[G, [G, G]]$ are both finite groups. Therefore there exists positive $n \in \mathbf{Z}$ such that $n[K] \in [G, [G, G]]$.

Choose positive $j, p \in \mathbf{Z}$ such that p and jn are relatively prime. Let $L \subset \partial N = \partial(\mathrm{nbd}(K))$ be a simple closed curve representing the homology class $(p)m + (jn)k$. Let $A \subset \partial(N)$ be an annulus neighborhood of L. Consider the manifold M' obtained via i left-handed Dehn twists along A. Since $A \subset \partial N$, M' is equivalent to some Dehn surgery on N, namely $M' = N(1+ijnp, ij^2n^2)$. On the other hand, after an isotopy A is contained in $\partial(\mathrm{nbd}(L))$, and so is equivalent to some Dehn surgery on L. Since $[L] = jn[K]$, $[L] \in [G, [G, G]]$. Therefore, by (4.9),

$$\delta(1 + ijnp, ij^2n^2) = \delta_0$$

for all i. Comparing this with (4.8), we get

$$\frac{ijnp}{dh(1+ijnp)}r([\frac{ij^2n^2}{1+ijnp}]) + \frac{ij^2n^2}{1+ijnp}\delta' = 0$$

(for $i \geq 0$).

Letting i, j and p take all possible values provides enough independent equations to prove that $\delta' = 0$ and $r(s) = 0$ for all $s \in \mathbf{Z}$. This completes the proof of (4.2), modulo that of (4.9). □

To prove (4.9), we will need the following lemma.

(4.10) **Lemma.** *Let $K^-, K^+ \subset M$ be two null-homologous knots in a $\mathbf{Q}HS$ which differ by a crossing change. (That is, K^- and K^+ coincide except in a 3-ball, where they differ as shown in Figure 3.3.) Let N^- and N^+ be the corresponding knot complements. Via the standard bases, we can identify $H_1(N^-;\mathbf{Z})$ with $H_1(N^+;\mathbf{Z})$ (see (1.G)). Then for any primitive $a \in H_1(N^-;\mathbf{Z}) = H_1(N^+;\mathbf{Z})$ not colinear with the longitude,*

$$\delta(N_a^-) = \delta(N_a^+).$$

□

The proof will be given later.

Proof of (4.9): By (4.10), it suffices to establish (4.9) for some knot in the homotopy class of K. That is, we are free to change crossings of K whenever it is convenient.

Since $[K] \in [\pi_1(M), [\pi_1(M), \pi_1(M)]]$, we may assume (after possibly changing some crossings of K) that K bounds an embedded surface $E \subset M$ which has a standard symplectic set of curves $X_1, Y_1, \ldots, X_k, Y_k \subset E$ such that each Y_i is null-homologous (i.e. $[Y_i] \in [\pi_1(M), \pi_1(M)]$). E can be thought of as a disk with bands attached, one band for each X_i or Y_i (see Figure 4.1). Let b_1 be the band associated to Y_1. Via further crossing changes, it can be arranged that the self-linking of b_1 is zero (see Figure 4.2).

Let $A \subset M$ be a curve parallel to Y_1, disjoint from E, and such that $\mathrm{lk}(Y_1, A) = 1$. Further require that A is close to Y_1, in the sense that an annulus bounded by A and Y_1 intersects E in two arcs contained in b_1 (see Figure 4.3). Let B be an unknotted curve linking Y_1 and disjoint from E, as shown in Figure 4.3.

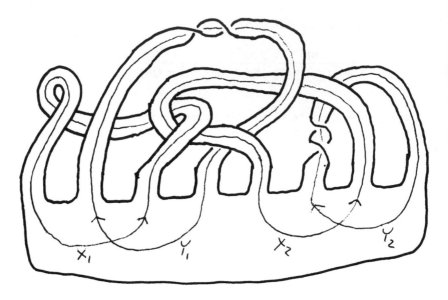

Figure 4.1: The surface E.

Figure 4.2: Changing the self linking of b_1.

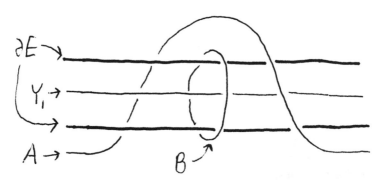

Figure 4.3: A, B, Y_1 and E.

Figure 4.4: L, surgery along which effects a crossing change.

A and B together bound a surface F which is disjoint from b_1. (This uses the fact that Y_1 is null-homologous and b_1 has self-linking number zero.) Without loss of generality, all intersections of F with E consist of bands of E passing though F. After still more crossing changes on K (i.e. passing bands through b_1), we may assume that F is disjoint from E and that A is still close to Y_1 in the above sense.

We are now in a position to apply (3.13) and (3.59). Let $M^\dagger$ denote M after $+1$ Dehn surgery on A and -1 Dehn surgery on B. Let $K^\dagger \subset M^\dagger$ denote the knot corresponding to K. By (3.13), $\lambda'(K^\dagger) = \lambda'(K)$. By (3.59), the function r (of (3.36)) also does not change. The double surgery along A and B does not change the linking numbers of any curves disjoint from F. Therefore, by (B.10), $\lambda'_c(K^\dagger) = \lambda'_c(K)$ (i.e. $\Gamma(K^\dagger) = \Gamma(K)$). So by (4.8), it suffices to prove (4.3) for $K^\dagger$.

Note, however, that in $M^\dagger$ E can be compressed along Y_1, reducing its genus by one. (The compressing disk passes once through each of the surgery solid tori of A and B.) Repeating the above procedure reduces our task to proving (4.3) for a genus zero knot (i.e. an unknot) in a **QHS**. This follows from (4.8), (3.51), and the obvious facts that $\lambda'(U) = \lambda'_c(U) = 0$ for an unknot U. □

Proof of (4.10): Let L be an unknot surrounding the crossing (see Figure 4.4), so that K_+ is equivalent to K_- in the manifold obtained by $+1$ surgery on L. (This manifold is, of course, diffeomorphic to M.) Note that $\mathrm{lk}(K_-, L) = 0$, so, by (3.59), surgery on L does not change the function r (of (3.36)). Therefore, by (4.8), what must be shown is that

$$
\begin{aligned}
0 &= \delta'(K_+) - \delta'(K_-) \\
&= \lambda'(K_+) - \lambda'(K_-) - \lambda'_c(K_+) + \lambda'_c(K_-).
\end{aligned}
$$

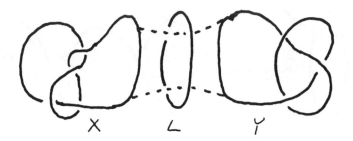

Figure 4.5: Smoothing the crossing, yielding X and Y.

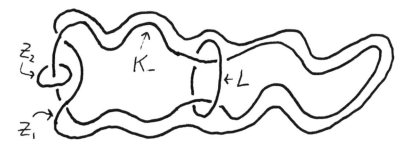

Figure 4.6: Z_1, Z_2, K_- and L

Let $X, Y \subset M$ be the curves obtained from "smoothing" the crossing, as shown in Figure 4.5. Note that K_- is a band connected sum of X and Y.

Using (B.10) it is an elementary exercise, left to the reader, to show that

$$\lambda_c'(K_+) - \lambda_c'(K_-) = \Gamma(K_+) - \Gamma(K_-) = -2\,\mathrm{lk}\,(X, Y).$$

(Hint: Select a Seifert surface for K_- which is disjoint from L, and select as part of its symplectic basis a curve which passes through L once and a curve which is parallel to X (or Y).) So the proof will be complete if it can be shown that

(4.11) $$\lambda''(K_-, L) = \lambda'(K_+) - \lambda'(K_-) = -2\,\mathrm{lk}\,(X, Y).$$

Let D be an unknotting disk for L which intersects K_- in two points. Let Z_1 be parallel to K_-, have linking number 1 with K_-, and be disjoint from D (see Figure 4.6). Let Z_2 be an unknot linking K_- once, as shown in Figure 4.6. Z_1 and Z_2 together bound a surface F which is disjoint from K_- and D. (F can be constructed by taking a Seifert surface for K_- which

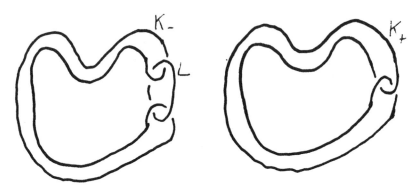

Figure 4.7: K_- and K_+ in $M^\dagger$

is disjoint from L, isotoping it so that it is disjoint from D and intersects K_- once transversely, and removing a neighborhood of the intersection point.) L bounds a genus one surface consisting of D with a neighborhood of $D \cap K_-$ excised and a tube surrounding half of K_-. This surface is disjoint from K_- and F.

We now apply (3.13). Let $M^\dagger$ be the manifold obtained from $+1$ surgery on Z_1 and -1 surgery on Z_2. By (3.13), this surgery does not change $\lambda'(L)$ (which is clearly zero anyway, since L is an unknot in both M and $M^\dagger$). The same is true (except for the parenthetical remark) if $+1$ surgery is done on K_- in both M and $M^\dagger$. Therefore the surgery on Z_1 and Z_2 does not change $\lambda''(K_-, L)$ (cf. (3.9)). It is easy to check that the surgery also does not change $\mathrm{lk}(X, Y)$. Thus it suffices to show that (4.11) holds in $M^\dagger$.

Note that in $M^\dagger$, X (say) can be isotoped through the surgery solid tori of Z_2 and Z_1 so that it ends up parallel to Y. It follows that in $M^\dagger$ K_- is unknotted and K_+ is a double, with negative clasp, of Y (see Figure 4.7). Hence $\lambda''(K_-, L) = \lambda'(K_+)$.

To each doubled knot K there is associated the self linking number of the core of K. In the above case this linking number is equal to $\mathrm{lk}(X, Y)$. So what must be shown is

(4.12) *Let $K \subset M$ be a doubled knot with negative clasp. Let $\mathrm{lk}(K) \in \mathbb{Q}$ denote the self linking number of the core of K. Then $\lambda'(K) = -2\mathrm{lk}(K)$.*

To prove this we will need

(4.13) **Lemma.** *Let V be a solid torus in a 3-manifold M. Let $(L, K) \subset V \subset M$ be the link shown in Figure 4.8. Then*

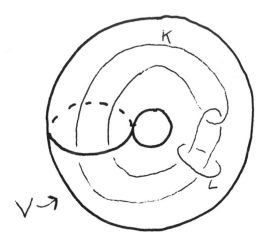

Figure 4.8: A two component link.

$$\lambda''(K, L) = 2.$$

Proof: Let $(Z_1, Z_2) \subset V$ be the link shown in Figure 4.9. L and the pair (Z_1, Z_2) bound disjoint Seifert surfaces in the complement of K (see Figure 4.10). Therefore, by (3.13), performing $(1, -1)$-surgery on (Z_1, Z_2) does not change $\lambda''(K, L)$. After this surgery is performed, we can slide K over Z_1, obtaining a two component link contained in a 3-ball, as shown in Figure 4.11. It now follows from Figure 4.12, (3.22) and (3.50) that $\lambda''(K, L) = 2$. $\square$

The strategy for proving (4.12) is to reduce to the case of an untwisted double $(\operatorname{lk}(K) = 0)$ and apply (3.26).

K bounds a genus one Seifert surface consisting of an annulus B parallel to the core of K and an "overpass" with one twist (see Figure 4.13). Let V be a solid torus neighborhood of the core of K such that K is contained in the interior of V. Let β be the core of the annulus B. Without loss of generality, ∂V contains an annulus neighborhood of β in B. Let $\mu \subset \partial V$ be a meridian of V. Let $m = [\mu] \in H_1(\partial V; \mathbf{Z})$ and $b = [\beta] \in H_1(\partial V; \mathbf{Z})$. Choose orientations of μ and β so that $\langle m, b \rangle = 1$. Let $l = pm + qb \in H_1(\partial V; \mathbf{Z})$ be a longitude of $\overline{M \setminus V}$. Clearly,

$$\operatorname{lk}(K) = -p/q.$$

Consider the effect of a left-handed Dehn twist along μ. Let $M^\dagger$ denote the new **QHS** so obtained, and let $K^\dagger \subset M^\dagger$ correspond to K. ($M^\dagger$ is, of

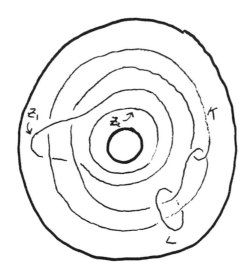

Figure 4.9: Another two component link.

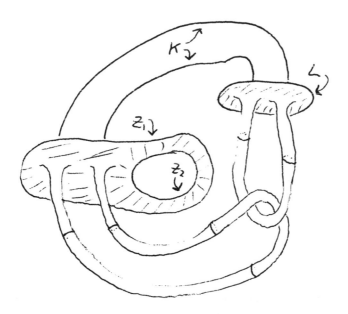

Figure 4.10: Seifert surfaces for L and (Z_1, Z_2).

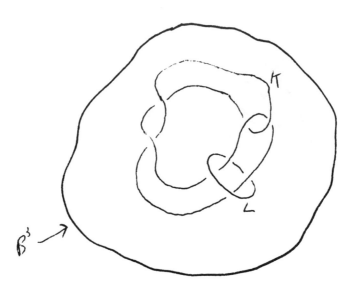

Figure 4.11: K and L after the surgery.

course, diffeomorphic to M.) $M^\dagger$ can also be thought of as the result of $+1$ Dehn surgery on μ. By (4.13),

$$\lambda'(K^\dagger) - \lambda'(K) \;=\; \lambda''(K, \mu)$$
(4.14)
$$=\; 2.$$

Let $l^\dagger = p^\dagger m + q^\dagger b$ be the longitude of $\overline{M^\dagger \setminus V}$. It is easy to see that

(4.15)
$$\begin{cases} p^\dagger = p + q \\ q^\dagger = q. \end{cases}$$

Hence $\mathrm{lk}\,(K^\dagger) = -(p+q)/q$, and so

(4.16)
$$- 2\mathrm{lk}\,(K^\dagger) + 2\mathrm{lk}\,(K) = 2.$$

It follows from (4.14) and (4.16) that (4.12) holds for K if and only if it holds for $K^\dagger$.

Now consider the effect of a right-handed Dehn twist along β. Let $M^\dagger$ denote the resulting 3-manifold, $K^\dagger \subset M^\dagger$ denote the knot corresponding to K, and $l^\dagger = p^\dagger m + q^\dagger b$ be the longitude of $\overline{M^\dagger \setminus V}$. It is easy to see that

(4.17)
$$\begin{cases} p^\dagger = p \\ q^\dagger = q - p. \end{cases}$$

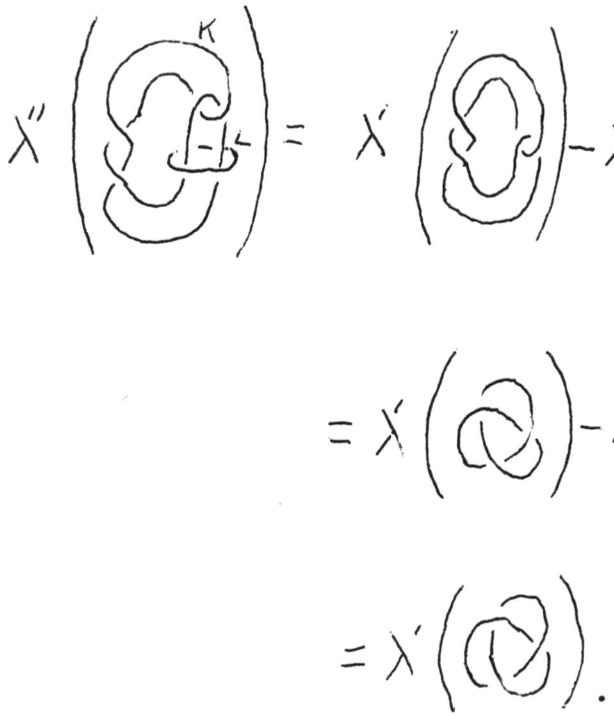

Figure 4.12: The end of the proof.

Figure 4.13: The Seifert surface for K

Assume that $q^\dagger$ has the same sign as q. In this case we can apply (3.16) and conclude

$$\bar{\lambda}'(K^\dagger) = \bar{\lambda}'(K).$$

Let

$$h = |\mathrm{Tor}(H_1(M \setminus V; \mathbf{Z}))|.$$

Then $|H_1(M; \mathbf{Z})| = h\langle m, l\rangle$ and $|H_1(M^\dagger; \mathbf{Z})| = h\langle m, l^\dagger\rangle$. Therefore

$$
\begin{aligned}
\lambda'(K^\dagger) &= \frac{\bar{\lambda}'(K^\dagger)}{|H_1(M^\dagger; \mathbf{Z})|} \\
&= \frac{|H_1(M; \mathbf{Z})|}{|H_1(M^\dagger; \mathbf{Z})|} \frac{\bar{\lambda}'(K)}{|H_1(M; \mathbf{Z})|} \\
(4.18) \qquad &= \frac{q}{q - p}\lambda'(K).
\end{aligned}
$$

On the other hand,

$$
\begin{aligned}
\mathrm{lk}(K^\dagger) &= -\frac{p}{q - p} \\
&= -\frac{q}{q - p}\frac{p}{q} \\
(4.19) \qquad &= \frac{q}{q - p}\mathrm{lk}(K).
\end{aligned}
$$

It follows from (4.18) and (4.19) that (4.12) holds for K if and only if it hold for $K^\dagger$.

Using the transformations (4.15) and (4.17) and their inverses (with the restriction that $q^\dagger$ and q have the same sign in (4.17)), one can reduce (4.12) to the case where $p = 0$. That is, $\mathrm{lk}(K) = 0$ and, by (3.26), $\lambda'(K) = 0$. This completes the proof of (4.12), and also that of (4.10). $\square$

§ 5

Combinatorial Definition of λ

In this section we show that the Dehn surgery formula of (4.2) (and the fact that $\lambda(S^3) = 0$) not only uniquely determine λ, but also can be used to give an alternative definition of λ.

For the sake of self-containedness, recall that if N is a 3-manifold, $\partial N \cong T^2$, and $a \in H_1(\partial N; \mathbf{Z})$ is primitive (represented by a simple closed curve), then N_a denotes the closed 3-manifold obtained from Dehn surgering N along a. (See also (1.G).) If N is the complement of a knot in a **QHS**, a *longitude* of N is defined to be a generator of $\ker(H_1(\partial N; \mathbf{Z}) \to H_1(N; \mathbf{Z}))$. This is unique up to sign.

All boundaries of manifolds are oriented according to the "inward normal last" convention.

For the definitions of τ and Γ, see the beginning of Section 4 or Sections A and B.

(5.1) **Theorem.** *There is a unique function*

$$\lambda : \{\text{rational homology spheres}\} \to \mathbf{Q}$$

such that

1. $\lambda(S^3) = 0$

2. *For N a knot complement is a **QHS** and l a longitude of N*

$$\lambda(N_b) = \lambda(N_a) + \tau(a, b; l) + \frac{\langle a, b \rangle}{\langle a, l \rangle \langle b, l \rangle} \Gamma(N)$$

for all primitive $a, b \in H_1(\partial N; \mathbf{Z}), \langle a, l \rangle \neq 0, \langle b, l \rangle \neq 0.$

(Note: A version of this theorem, applying to the case where both N_a and N_b are homology lens spaces, has been proved independently by Boyer and Lines [BL].)

The proof of (5.1) occupies the rest of this section. Roughly, the idea of the proof is to represent a rational homology sphere M as surgery on a link in S^3, compute $\lambda(M)$ by surgering one component at a time and applying (5.1.2) (and (5.1.1)), and then show that the result is independent of the choice of surgery representation.

(5.2) Definition. A *permissible surgery sequence* (or PSS) $\mathcal{N} = (N^i,\, m_i,\, s_i,\, h_i)$ consists of 3-manifolds N^i with $\partial N_i \cong T^2$ $(1 \le i \le n)$, primitive elements $m_i, s_i \in H_1(\partial N^i; \mathbf{Z})$ $(1 \le i \le n)$, and homeomorphisms $h_i : N^i_{s_i} \to N^{i+1}_{m_{i+1}}$ $(1 \le i \le n-1)$ such that $N^1_{m_1} \cong S^3$ and each $N^i_{s_i}$ is a QHS. We say $\mathcal{N}$ *represents* $N^n_{s_n}$. Two PSS's $\mathcal{N} = (N^i, m_i, s_i, h_i)$ and $\mathcal{N}' = (N'^i, m'_i, s'_i, h'_i)$ are considered the same if there is a commutative ladder

$$
\begin{array}{ccccccccc}
\cdots & \hookleftarrow & N^i & \hookleftarrow & N^i_{s_i} & \xrightarrow{h_i} & N^{i+1}_{m_{i+1}} & \hookleftarrow & N^{i+1} & \hookleftarrow & \cdots \\
 & & \downarrow & & \downarrow & & \downarrow & & \downarrow & & \\
\cdots & \hookleftarrow & N'^i & \hookleftarrow & N'^i_{s'_i} & \xrightarrow{h'_i} & N'^{i+1}_{m'_{i+1}} & \hookleftarrow & N'^{i+1} & \hookleftarrow & \cdots
\end{array}
$$

which takes m_i to m'_i and s_i to s'_i.

Given a PSS $\mathcal{N}$, define $\lambda(\mathcal{N})$ to be the value of $\lambda(N^n_{s_n})$ computed by applying (5.1.1) and (5.1.2) to the sequence of surgeries in $\mathcal{N}$. That is,

$$(5.3) \qquad \lambda(\mathcal{N}) \stackrel{\text{def}}{=} \sum_{i=1}^{n} \left(\tau(m_i, s_i; l_i) + \frac{\langle m_i, s_i \rangle}{\langle m_i, l_i \rangle \langle s_i, l_i \rangle} \Gamma(N^i) \right),$$

where $l_i \in H_1(\partial N^i; \mathbf{Z})$ is a longitude of N^i.

The unicity of λ follows from

(5.4) Lemma. *Every* **Q**HS *is represented by a* PSS. $\qquad\qquad\square$

The proof will be given later.

(Note that if the requirement of rationality is dropped, this is equivalent to the fact ([L]) that every closed 3-manifold can be obtained from Dehn surgery on a link in S^3.)

The proof will be complete if we can show that $\lambda(\mathcal{N}) = \lambda(\mathcal{N}')$ for any two PSS's representing the same **Q**HS M, for then we can (well-) define

$\lambda(M)$ to be $\lambda(\mathcal{N})$ (for any $\mathcal{N}$ representing M), and $\lambda(M)$ so defined clearly satisfies (5.1.1) and (5.1.2).

(5.5) We now introduce four "moves" for modifying PSS's. Throughout, $\mathcal{N}$ will denote (N^i, m_i, s_i, h_i), $1 \leq i \leq n$.

1. *Isotopy.* Change the homeomorphisms h_i by isotopies.

2. *Stabilization.* Fix $0 \leq j \leq n$ and let $K \subset N$ be an unknotted solid torus in $N_{s_j}^j$. (If $j = 0$, define $N_{s_j}^j$ to be $N_{m_1}^1$.) Let $N' = N_{s_j}^j \backslash K$, m' be a meridian of K, and $s' = m' + l'$, where l' is a longitude of N'. Let $h' : N_{s'}' \rightarrow N_{s_{j+1}}^{j+1}$ be a homeomorphism which coincides with h_j outside a regular neighborhood of K union its unknotting disk. Insert (N', m', s', h') into $\mathcal{N}$ after (N^j, m_j, s_j, h_j).

3. *Prolongation.* Fix j and let r be a primitive element of $H_1(\partial N^j; \mathbf{Z})$ such that $\langle r, l \rangle \neq 0$ (l a longitude of N^j). Replace (N^j, m_j, s_j, h_j) with $(N^j, m_j, r, \mathrm{id})$, (N^j, r, s_j, h_j).

4. *Permissible transposition.* Fix $1 \leq j \leq n - 1$. Suppose that $\overline{N_{s_j}^j \backslash N^j}$ is disjoint from $h_j^{-1}(\overline{N_{m_{j+1}}^{j+1} \backslash N^{j+1}})$, that is, the jth and $j + 1^{\mathrm{st}}$ surgery tori are disjoint. Let $R \overset{\mathrm{def}}{=} N^j \cap h_j^{-1}(N^{j+1})$. We can identify ∂R with $\partial N^j \cup \partial N^{j+1}$. We have the following commutative diagram :

$$
\begin{array}{ccccc}
R_{m_j m_{j+1}} & \leftarrow & R_{m_{j+1}} & \rightarrow & R_{s_j m_{j+1}} \\
\uparrow & & \uparrow & & \uparrow \\
R_{m_j} & \leftarrow & R & \rightarrow & R_{s_j} \\
\downarrow & & \downarrow & & \downarrow \\
R_{m_j s_{j+1}} & \leftarrow & R_{s_{j+1}} & \rightarrow & R_{s_j s_{j+1}}.
\end{array}
$$

All maps are inclusions. Note that $R_{m_{j+1}} = N^j$, $R_{s_j} = N^{j+1}$, $R_{s_j m_{j+1}} = N_{s_j}^j = N_{m_{j+1}}^{j+1}$, etc. Suppose that the closed 3-manifold $R_{m_j s_{j+1}}$ is a **Q**HS. Then we can form a new PSS $\mathcal{N}'$ by replacing N^j with R_{m_j}, N^{j+1} with $R_{s_j s_{j+1}}$ and interchanging m_j, s_j and m_{j+1}, s_{j+1}. In other words, we interchange the order of the jth and $j + 1^{\mathrm{st}}$ surgeries.

(5.6) Lemma. *Any two PSS's which represent the same **Q**HS are related by a finite sequence of isotopies, stabilizations (and inverse stabilizations), prolongations and permissible transpositions.* □

The proof will be given later.

   (If the requirement of rationality is dropped, this follows easily from [K1].)

Clearly $\lambda(\mathcal{N})$ does not change if we modify $\mathcal{N}$ by an isotopy.

If $\mathcal{N}$ is changed by a stabilization, then (using the notation from the definition of stabilization) $\lambda(\mathcal{N})$ changes by

$$\tau(m', m' + l'; l') + \frac{\langle m', m' + l' \rangle}{\langle m', l' \rangle \langle m' + l', l' \rangle} \Gamma(N').$$

The first term is zero by (A.15) and (A.13). The second term is zero because N' is the complement of an unknot, so $\Gamma(N') = 0$. Thus stabilizations do not change $\lambda(\mathcal{N})$.

If $\mathcal{N}$ is modified by a prolongation, (retaining the above notation and dropping the "j" 's) $\lambda(\mathcal{N})$ changes by

$$\tau(m, r; l) + \tau(r, s; l) - \tau(m, s; l)$$
$$+ \frac{\langle m, r \rangle}{\langle m, l \rangle \langle r, l \rangle} \Gamma(N) + \frac{\langle r, s \rangle}{\langle r, l \rangle \langle s, l \rangle} \Gamma(N) - \frac{\langle m, s \rangle}{\langle m, l \rangle \langle s, l \rangle} \Gamma(N).$$

The τ terms cancel by (A.14). The coefficients of $\Gamma(N)$ add up to zero by elementary linear algebra. Hence prolongations do not change $\lambda(\mathcal{N})$.

Thus we have reduced (5.1) to

(5.7) **Lemma.** *Let $\mathcal{N}$ and $\mathcal{N}'$ be two PSS's which differ by a permissible transposition. Then $\lambda(\mathcal{N}) = \lambda(\mathcal{N}')$.*

Proof: Recall the notation from the definition of permissible transposition. Let $a = m_j$, $b = s_j$, $x = m_{j+1}$ and $y = s_{j+1}$. Let $\partial_1 = \partial R_x = \partial R_y \subset \partial R$ and $\partial_2 = \partial R_a = \partial R_b \subset \partial R$. Thus $a, b \in H_1(\partial_1; \mathbf{Z})$ and $x, y \in H_1(\partial_2; \mathbf{Z})$. Let l_1, l_1', l_2 and l_2' be longitudes of R_x, R_y, R_a and R_b, respectively. In view of (5.3), what we must show is

(5.8) $\tau(a, b; l_1) + \tau(x, y; l_2') + \dfrac{\langle a, b \rangle}{\langle a, l_1 \rangle \langle b, l_1 \rangle} \Gamma(R_x) + \dfrac{\langle x, y \rangle}{\langle x, l_2' \rangle \langle y, l_2' \rangle} \Gamma(R_b) =$

$\tau(x, y; l_2) + \tau(a, b; l_1') + \dfrac{\langle x, y \rangle}{\langle x, l_2 \rangle \langle y, l_2 \rangle} \Gamma(R_a) + \dfrac{\langle a, b \rangle}{\langle a, l_1' \rangle \langle b, l_1' \rangle} \Gamma(R_y).$

Consider the map $i_* : H_1(\partial_1; \mathbf{Q}) \to H_1(R; \mathbf{Q})$. i_* must either be an isomorphism or have rank 1.

Case (i). i_* an isomorphism. In this case the inclusions induce isomorphisms

$$H_1(\partial_1; \mathbf{Q}) \cong H_1(R; \mathbf{Q}) \cong H_1(\partial_2; \mathbf{Q}).$$

We will identify these spaces via the above isomorphisms and call them collectively V. Let $\langle \cdot, \cdot \rangle_i$ denote the intersection pairing on $H_1(\partial_i; \mathbf{Q})$. Note that under this identification $\langle \cdot, \cdot \rangle_1 = -\langle \cdot, \cdot \rangle_2$. Let $\langle \cdot, \cdot \rangle \overset{\text{def}}{=} \langle \cdot, \cdot \rangle_1$. This inner product allows us to identify V and V^*. V has four distinguished lattices: $H_1(\partial_1; \mathbf{Z})$, $H_1(\partial_2; \mathbf{Z})$, Free$(H_1(R; \mathbf{Z}))$ and Hom$(H_1(R; \mathbf{Z}), \mathbf{Z}) \cong H_2(R, \partial R; \mathbf{Z})$. For $r \in V$, define $\nu(r)$ to be the minimal positive integer such that $\nu(r)r \in H_2(R, \partial R; \mathbf{Z})$.

We will need the following two lemmas.

(5.9) **Lemma.** *There is a symmetric bilinear form B defined on V such that*

$$\Gamma(R_{r_i}) = \nu(r_i)^2 B(r_i, r_i) - \frac{1}{12}\nu(r_i)^2 + \frac{1}{12}$$

for all primitive $r_i \in H_1(\partial_i; \mathbf{Z})$, $i = 1$ or 2.

Proof: Let $\widetilde{R_{r_i}}$ be the universal free abelian cover of R_{r_i}. Let $\tilde{R}$ be the induced cover of $R \subset R_{r_i}$. Let $\tilde{R}'$ be the universal free abelian cover of R. We have the following commutative diagram

$$
\begin{array}{ccc}
\tilde{R}' & & \\
\downarrow & & \\
\tilde{R} & \hookrightarrow & \widetilde{R_{r_i}} \\
\downarrow & & \downarrow \\
R & \hookrightarrow & R_{r_i}.
\end{array}
$$

Let

$$\Delta(\widetilde{R_{r_i}}), \Delta(\tilde{R}) \in \mathbf{Q}[\text{Free}(H_1(R_{r_i}; \mathbf{Z}))]$$
$$\Delta(\tilde{R}') \in \mathbf{Q}[\text{Free}(H_1(R; \mathbf{Z}))]$$

be the Alexander polynomials. We will express $\Delta(\tilde{R})$ in terms of $\Delta(\tilde{R}')$ and then $\Delta(\widetilde{R_{r_i}})$ in terms of $\Delta(\tilde{R})$.

The inclusion $i : R \hookrightarrow R_{r_i}$ induces a map

$$i_* : \mathbf{Q}[\text{Free}(H_1(R; \mathbf{Z}))] \rightarrow \mathbf{Q}[\text{Free}(H_1(R_{r_i}; \mathbf{Z}))].$$

Let t be a generator of Free$(H_1(R_{r_i}; \mathbf{Z}))$, written multiplicatively. By (B.4),

$$\Delta(\tilde{R}) = (1 - t)i_*(\Delta(\tilde{R}')).$$

More concretely, recall that $\mathrm{Free}\,(H_1(R;\mathbf{Z}))$ can be identified with a lattice in V. Then for $x \in \mathrm{Free}\,(H_1(R;\mathbf{Z}))$,

$$i_*(x) = t^{\langle x, \nu(r_i) r_i \rangle}$$

(after possibly replacing t with t^{-1}). Thus, if $\Delta(\widetilde{R}') = \sum_j c_j [x_j]$ ($c_j \in \mathbf{Q}$, $x_j \in \mathrm{Free}\,(H_1(R;\mathbf{Z}))$), then

$$(5.10) \qquad \Delta(\widetilde{R}) = (1-t) \sum_j c_j t^{\langle x_j, \nu(r_i) r_i \rangle}.$$

We now relate $\Delta(\widetilde{R_{r_i}})$ to $\Delta(\widetilde{R})$. The map

$$i_* : H_1(\widetilde{R};\mathbf{Q}) \rightarrow H_1(\widetilde{R_{r_i}};\mathbf{Q})$$

is a surjective map of $\mathbf{Q}[\mathrm{Free}\,(H_1(R_{r_i};\mathbf{Z}))]$ modules. Therefore, by (B.1),

$$(5.11) \qquad \Delta(\widetilde{R}) = \Delta(\widetilde{R_{r_i}})\Delta(\ker(i_*)).$$

Let $\partial_k = \partial R \backslash \partial R_{r_i}$. The inverse image $\widetilde{\partial_k}$ of ∂_k in $\widetilde{R}'$ consists of $\nu(r_i)$ copies of $S^1 \times \mathbf{R}$. It is not hard to see that the inclusion induced map $H_1(\widetilde{\partial_k};\mathbf{Q}) \rightarrow H_1(\widetilde{R};\mathbf{Q})$ is an isomorphism onto $\ker(i_*)$. $\mathrm{Free}\,(H_1(R_{r_i};\mathbf{Z}))$ (acting as deck translations) cyclically permutes the components of $\widetilde{\partial_k}$. It follows that

$$\Delta(\ker(i_*)) = t^{\nu(r_i)} - 1.$$

Combining this with (5.10) and (5.11) yields

$$(5.12) \qquad \Delta(\widetilde{R_{r_i}}) = \frac{\sum c_j t^{\langle x_j, \nu(r_i) r_i \rangle}}{1 + t + \cdots + t^{\nu(r_i)-1}}.$$

In order to compute $\Gamma(R_{r_i})$, we must normalize and symmetrize (5.12). It follows from the results of [M2] that we may assume that $\Delta(\widetilde{R}') = \sum c_j [x_j]$ is symmetric under $x_j \leftrightarrow -x_j$, though possibly at the cost of $x_j \in \frac{1}{2}\mathrm{Free}\,(H_1(R;\mathbf{Z})) \subset V$. This implies that the numerator of (5.12) is symmetric under $t \leftrightarrow t^{-1}$. We may further assume that $\sum c_j = 1$. Let

$$E_l(t) \overset{\text{def}}{=} l^{-1} \sum_{j=-\frac{l-1}{2}}^{j=\frac{l-1}{2}} t^j.$$

$E_{\nu(r_i)}$ is the normal, symmetric form of the denominator of (5.12). Hence the normal, symmetric form of the Alexander polynomial of R_{r_i} is

$$\Delta_{\mathrm{SN}}(R_{r_i}) \overset{\text{def}}{=} \frac{\sum c_j t^{\langle x_j, \nu(r_i) r_i \rangle}}{E_{\nu(r_i)}(t)}.$$

Taking the formal second derivative with respect to t and evaluating at $t = 1$, we have

$$\Gamma(R_{r_i}) = \frac{d^2}{dt^2}\Delta_{\text{SN}}(R_{r_i})|_{t=1} = \nu(r_i)^2 B(r_i, r_i) - \frac{1}{12}\nu(r_i)^2 + \frac{1}{12}$$

where

$$B(a, b) \stackrel{\text{def}}{=} \sum c_j \langle x_j, a\rangle\langle x_j, b\rangle.$$

$\square$

(5.13) **Lemma.** *Let C be any symmetric bilinear form on V, and let a_1, b_1, a_2, $b_2 \in V$ be pairwise linearly independent. Then*

$$\frac{\langle a_2, b_2\rangle}{\langle a_2, a_1\rangle\langle b_2, a_1\rangle}C(a_1, a_1) - \frac{\langle a_2, b_2\rangle}{\langle a_2, b_1\rangle\langle b_2, b_1\rangle}C(b_1, b_1)$$

$$+\frac{\langle a_1, b_1\rangle}{\langle a_1, a_2\rangle\langle b_1, a_2\rangle}C(a_2, a_2) - \frac{\langle a_1, b_1\rangle}{\langle a_1, b_2\rangle\langle b_1, b_2\rangle}C(b_2, b_2) \;\;=\;\; 0.$$

Proof: There exist non-zero $\alpha, \beta, \gamma, \delta \in \mathbf{Q}$ such that $a_1 = \alpha a_2 + \beta b_2$ and $b_1 = \gamma a_2 + \delta b_2$. Hence

$$\begin{aligned}
C(a_1, a_1) &= \alpha^2 C(a_2, a_2) + \alpha\beta C(a_2, b_2) + \beta^2 C(b_2, b_2)\\
C(b_1, b_1) &= \gamma^2 C(a_2, a_2) + \gamma\delta C(a_2, b_2) + \delta^2 C(b_2, b_2).
\end{aligned}$$

Eliminating $C(a_2, b_2)$ gives

$$\frac{1}{\alpha\beta}C(a_1, a_1) - \frac{1}{\gamma\delta}C(b_1, b_1) = (\frac{\alpha}{\beta} - \frac{\gamma}{\delta})C(a_2, a_2) + (\frac{\beta}{\alpha} - \frac{\delta}{\gamma})C(b_2, b_2).$$

Substituting $\alpha = \langle a_1, b_2\rangle/\langle a_2, b_2\rangle$, $\beta = \langle a_1, a_2\rangle/\langle b_2, a_2\rangle$, etc., and noting that $\alpha\delta - \beta\gamma = \langle a_1, b_1\rangle/\langle a_2, b_2\rangle$, we get

$$\frac{\langle a_2, b_2\rangle\langle b_2, a_2\rangle}{\langle a_1, b_2\rangle\langle a_1, a_2\rangle}C(a_1, a_1) - \frac{\langle a_2, b_2\rangle\langle b_2, a_2\rangle}{\langle b_1, b_2\rangle\langle b_1, a_2\rangle}C(b_1, b_1) =$$

$$\frac{\langle a_2, b_2\rangle\langle a_1, b_1\rangle}{\langle a_1, a_2\rangle\langle b_1, a_2\rangle}C(a_2, a_2) - \frac{\langle a_2, b_2\rangle\langle a_1, b_1\rangle}{\langle a_1, b_2\rangle\langle b_1, b_2\rangle}C(b_2, b_2).$$

Dividing by $\langle a_2, b_2\rangle$ completes the proof. $\square$

We now establish (5.7) for case (i). First observe that $l_1 = \nu(x)x$, $l_1' = \nu(y)y$, $l_2 = \nu(a)a$ and $l_2' = \nu(b)b$. Also, since we have identified $H_1(\partial_2; \mathbf{Z})$ with a lattice in V and the pairing on V has the opposite sign from the intersection pairing on $H_1(\partial_2; \mathbf{Z})$, the signs of terms in (5.8) involving elements elements of $H_1(\partial_2; \mathbf{Z})$ must be changed. (5.8) now becomes

(5.14) $\tau(a, b; \nu(x)x) - \tau(a, b; \nu(y)y) + \tau(x, y; \nu(a)a) - \tau(x, y; \nu(b)b) =$

$$-\frac{\langle a, b \rangle}{\langle a, \nu(x)x \rangle \langle b, \nu(x)x \rangle} \Gamma(R_x) + \frac{\langle a, b \rangle}{\langle a, \nu(y)y \rangle \langle b, \nu(y)y \rangle} \Gamma(R_y)$$

$$-\frac{\langle x, y \rangle}{\langle x, \nu(a)a \rangle \langle y, \nu(a)a \rangle} \Gamma(R_a) + \frac{\langle x, y \rangle}{\langle x, \nu(b)b \rangle \langle y, \nu(b)b \rangle} \Gamma(R_b).$$

By (5.9) and (5.13), the right hand side is equal to

$$\frac{1}{12} \left(\frac{\langle a, b \rangle}{\langle a, x \rangle \langle b, x \rangle} - \frac{\langle a, b \rangle}{\langle a, y \rangle \langle b, y \rangle} + \frac{\langle x, y \rangle}{\langle x, a \rangle \langle y, a \rangle} - \frac{\langle x, y \rangle}{\langle x, b \rangle \langle y, b \rangle} \right.$$

$$-\frac{\langle a, b \rangle}{\langle a, \nu(x)x \rangle \langle b, \nu(x)x \rangle} + \frac{\langle a, b \rangle}{\langle a, \nu(y)y \rangle \langle b, \nu(y)y \rangle}$$

$$\left. -\frac{\langle x, y \rangle}{\langle x, \nu(a)a \rangle \langle y, \nu(a)a \rangle} + \frac{\langle x, y \rangle}{\langle x, \nu(b)b \rangle \langle y, \nu(b)b \rangle} \right).$$

By (A.16), this differs from the left hand side by

$$\frac{1}{4} \left(\text{sign} \frac{\langle a, b \rangle}{\langle a, x \rangle \langle b, x \rangle} - \text{sign} \frac{\langle a, b \rangle}{\langle a, y \rangle \langle b, y \rangle} \right.$$

$$\left. + \text{sign} \frac{\langle x, y \rangle}{\langle x, a \rangle \langle y, a \rangle} - \text{sign} \frac{\langle x, y \rangle}{\langle x, b \rangle \langle y, b \rangle} \right).$$

This is zero, and the proof of (5.7) is complete in case (i).

Case (ii). i_* has rank one. In this case, $l_1 = l_1'$ and $l_2 = l_2'$, so (5.8) reduces to

(5.15) $\dfrac{\langle a, b \rangle}{\langle a, l_1 \rangle \langle b, l_1 \rangle} (\Gamma(R_x) - \Gamma(R_y)) = \dfrac{\langle x, y \rangle}{\langle x, l_2 \rangle \langle y, l_2 \rangle} (\Gamma(R_a) - \Gamma(R_b)).$

This follows immediately from

(5.16) **Lemma.** *There is a $\beta = \beta(R) \in \mathbf{Q}$ such that*

$$\Gamma(R_a) - \Gamma(R_b) = 2 \frac{\langle a, b \rangle}{\langle a, l_1 \rangle \langle b, l_1 \rangle} \beta$$

$$\Gamma(R_x) - \Gamma(R_y) = 2 \frac{\langle x, y \rangle}{\langle x, l_2 \rangle \langle y, l_2 \rangle} \beta.$$

(If R is the complement of a two component link in S^3 with linking number zero, then β is the Sato-Levine invariant.)

Proof: We will use (B.10) to compute Γ. There are connected surfaces E_1, $E_2 \in R$ such that $[\partial E_i] = l_i \in H_1(\partial_i; \mathbf{Z})$. Put E_1 and E_2 in general position. Then $E_1 \cap E_2$ is a closed 1-manifold. Orient $E_1 \cap E_2$ so that its orientation followed by the positive normal of E_1 followed by the positive normal of E_2 gives the orientation of R. Let $(E_1 \cap E_2)^+$ denote the push-off of $E_1 \cap E_2$ in E_1 in the positive normal direction of E_2. Define

$$\beta(R) \overset{\text{def}}{=} \mathrm{lk}\,(E_1 \cap E_2, (E_1 \cap E_2)^+).$$

It is not hard to show that this does not depend on the choice of E_1 and E_2.

Let $\alpha_1, \ldots, \alpha_{2g}$ be curves representing a symplectic basis of $H_1(E_2; \mathbf{Z})$ and such that $[E_1 \cap E_2] = k[\alpha_2]$ for some $k \in \mathbf{Z}$. Since $\langle [\alpha_i], [E_1 \cap E_2] \rangle = 0$ for $i \geq 2$, $\alpha_2, \ldots, \alpha_{2g}$ rationally bound surfaces in R. Thus any linking number involving them is independent of how ∂_1 is surgered. It follows from this and (B.10) that

$$\Gamma(R_a) - \Gamma(R_b) = 2(\mathrm{lk}\,_a(\alpha_1^-, \alpha_1) - \mathrm{lk}\,_b(\alpha_1^-, \alpha_1))\mathrm{lk}\,(\alpha_2^+, \alpha_2),$$

where $\mathrm{lk}\,_a$ [$\mathrm{lk}\,_b$] denotes the linking number in R_a [R_b].

Let $N(\alpha_1)$ be a closed tubular neighborhood of α_1 in R and let $R^- \overset{\text{def}}{=} \overline{R \setminus N(\alpha_1)}$. As in case (i) above, the inclusions into R^- induce an isomorphism

$$H_1(\partial_1; \mathbf{Q}) \cong H_1(\partial N(\alpha_1); \mathbf{Q}),$$

and we identify these two spaces. Let $p = [\alpha_1^-] \in H_1(\partial N(\alpha_1); \mathbf{Q})$, and let $m \in H_1(\partial N(\alpha_1); \mathbf{Q})$ be a meridian of $N(\alpha_1)$ such that $\langle m, p \rangle = 1$. Then

$$\mathrm{lk}\,_c(\alpha_1^-, \alpha_1) = -\frac{\langle p, c \rangle}{\langle m, c \rangle},$$

where $c = a$ or b. Note also that $km = l_1$, up to sign. Therefore

$$
\begin{aligned}
\Gamma(R_a) - \Gamma(R_b) &= 2\left(-\frac{\langle p, a \rangle}{\langle m, a \rangle} + \frac{\langle p, b \rangle}{\langle m, b \rangle}\right) \mathrm{lk}\,(\alpha_2^+, \alpha_2) \\
&= 2\left(\frac{\langle p, a \rangle \langle m, b \rangle - \langle m, a \rangle \langle p, b \rangle}{\langle m, a \rangle \langle m, b \rangle}\right)\left(\frac{\beta(R)}{k^2}\right) \\
&= 2\frac{\langle a, b \rangle}{\langle a, l_1 \rangle \langle b, l_1 \rangle}\beta(R).
\end{aligned}
$$

This completes the proof of the first half of (5.16). The proof of the other half is similar. □

The proof of (5.7) is also complete. □

It remains to prove (5.4) and (5.6). The first step is to reformulate them in terms of framed link diagrams.

Definitions. An *ordered framed link* (or OFL) $(L, s) = ((L_1, s_1), \ldots, (L_n, s_n))$ consists of an ordered link $L = (L_1, \ldots, L_n) \subset S^3$ and primitive homology classes $s_i \in H_1(\partial(\mathrm{nbd}(L_i)); \mathbf{Z})$. Let m_i be a meridian of $\mathrm{nbd}(L_i)$ and l_i be a longitude of $S^3 \setminus \mathrm{nbd}(L_i)$, oriented so that $\langle m_i, l_i \rangle = 1$. Then $s_i = \alpha_i m_i + \beta_i l_i$ ($\alpha_i, \beta_i \in \mathbf{Z}$), and we identify s_i with $\alpha_i / \beta_i \in \mathbf{Q} \cup \{\infty\}$. The *result* of (L, s), $M(L, s)$, is $S^3 \setminus \mathrm{nbd}(L)$ surgered along $s_1, \ldots, s_n$. (L, s) is *permissible* if $M((L_1, s_1), \ldots, (L_k, s_k))$ is a QHS for each $k \leq n$. (L, s) is *integral* if $\langle m_i, s_i \rangle$ is an integer for each i, or, equivalently, if the rational number corresponding to s_i is an integer. An integral PSS is defined similarly.

Let (L, s) be an integral OFL. Choose orientations of the components of L and represent each s_i by a simple closed curve $S_i \subset \partial(\mathrm{nbd}(L_i))$ which is (oriented) parallel to L_i. The matrix $[\mathrm{lk}\,(L_i, S_j)]$ is called the *linking matrix* of (L, s). ($\mathrm{lk}\,(\cdot, \cdot)$ denotes the linking number in S^3.) It is not hard to show that the linking matrix is a presentation matrix for $H_1(M(L, s); \mathbf{Z})$. This implies

(5.17) Lemma. *An integral OFL (L, s) is permissible if and only if*

$$\mathrm{det}_k(L, s) \stackrel{\mathrm{def}}{=} \det([\mathrm{lk}\,(L_i, S_j)]_{1 \leq i, j \leq k})$$

is non-zero for each $k \leq n$. □

A permissible OFL determines a PSS in the obvious way: In the notation of (5.2), $N^i = M((L_1, s_1), \ldots, (L_{i-1}, s_{i-1})) \setminus \mathrm{nbd}(L_i)$. Modulo isotopies of PSS's, every PSS corresponds to some OFL.

(5.18) Recall ([K1]) that any two integral OFL's representing the same 3-manifold are related by a finite sequence of the following three moves.

[Inverse] stabilization. Add [subtract] an unknotted and unlinked component K to L with framing ±1. K may be inserted anywhere in the ordering.

Handle slides. Fix i and j. Let S_i be a simple closed curve in $\partial\,\mathrm{nbd}(L_i)$
   representing $\pm s_i$. Replace L_j with a band connected sum of L_j and
   S_i (use any band). If L_i and L_j have framings a_i and $a_j \in \mathbf{Z}$, the
   framing for the new L_j is $a_i + a_j \pm \mathrm{lk}\,(L_i, L_j)$. (The sign depends
   on whether the band preserves or reverses the orientations of L_i and
   L_j.) In terms of the linking matrix, this corresponds to adding or
   subtracting the ith row and column to the jth row and column.

Transpositions. Exchange (L_i, s_i) and (L_{i+1}, s_{i+1}).

   As a final preliminary, we need the following lemma, the proof of which
is left to the reader.

(5.19) **Lemma.** *Let* $P_1, \dots, P_m$ *be polynomials defined on* $\mathbf{Z}^k$. *If, for each*
$1 \le i \le m$, *there is a* $v_i \in \mathbf{Z}^k$ *such that* $P_i(v_i) \ne 0$, *then there is a* $v \in \mathbf{Z}^k$
such that $P_i(v) \ne 0$ *for all* $1 \le i \le m$. $\square$

   We are now ready to prove (5.4). In view of the above remarks, this
follows from

(5.20) **Lemma.** *Every* **Q***HS* M *is represented by a permissible (and also
integral) OFL.*

Proof: By [L], every 3-manifold is represented by an integral OFL. Let
(L, s) represent M. We will modify (L, s) by stabilizations and handle
slides to obtain a permissible integral OFL.
   First add unknots $K_1, \dots, K_n$ with framing 1 to (L, s) at the end of
the ordering. Next, choose integers $b_1, \dots, b_n$ and slide each L_i over K_i b_i
times. Call the resulting OFL (L', s'). Let A be the linking matrix of (L, s)
and B be the diagonal matrix with diagonal $(b_1, \dots, b_n)$. Then the linking
matrix of (L', s') is

$$A' = \begin{bmatrix} A + B^2 & B \\ B & I \end{bmatrix}.$$

   We must choose the b_i's so that $\det_k(L', s') \ne 0$ for all $k \le 2n$ (see
(5.17)). $\det_k(L', s')$ is a polynomial function of the b_i's. In view of (5.19),
it suffices to show that each of these polynomials individually has a nonso-
lution. If we choose the b_i's sufficiently large, the matrix $A + B^2$ (thought
of as a bilinear form) will be positive definite, and hence $\det_k(L', s')$ will

be positive for $k \leq n$. If the b_i's are all zero, then, for $n + 1 \leq k \leq 2n$, $\det_k(L', s') = \det(A) \neq 0$.   □

Next, as a step toward (5.6), we prove

(5.21) Lemma. *Any PSS can be prolonged (in the sense of (5.5.3)) to an integral PSS.*

Proof: Recall the notation of (5.2). It suffices to find, for each i, primitive homology classes $c_{i0} = m_i$, c_{i1}, ..., $c_{ik_i} = s_i \in H_1(\partial N^i; \mathbf{Z})$ such that $\langle c_{ij}, c_{ij+1} \rangle = \pm 1$ and $\langle c_{ij}, l_i \rangle \neq 0$ for all j. This is possible, by elementary number theory.   □

We now introduce moves analogous to those in (5.6) for modifying permissible integral OFL's.

Ordered handle slides. Slide L_i over L_j, *where* $i \geq j$. The corresponding PSS changes by an isotopy.

[Inverse] stabilizations. Same as in (5.18). The corresponding PSS changes by a stabilization (as defined in (5.5)).

Permissible transpositions. Interchange L_i and L_{i+1} in the ordering, provided that the resulting OFL is also permissible. The corresponding PSS changes by a permissible transposition (as defined in (5.5)).

The proof of (5.6) will be complete upon the proof of

(5.22) Lemma. *Any two permissible integral OFL's, (L, s) and (L', s'), which represent the same rational homology sphere are related by a finite sequence of ordered handle slides, [inverse] stabilizations and permissible transpositions.*

Proof: After stabilizing (L, s) and (L', s'), we may assume, by Kirby's theorem, that (L, s) and (L', s') are related by a finite sequence of ordered handle slides and transpositions. Unfortunately, the transpositions are not necessarily permissible. In order to correct this, we do the following.

Step 1.   Add n (= the number of components of L and L') unknots $K_1, \ldots, K_n$ with framing 1 to (L, s) at the beginning of the ordering.

Step 2. Choose integers $b_1, \ldots, b_n$, satisfying conditions specified below, and slide L_i over K_i b_i times.

Step 3. Via transpositions, move $K_1, \ldots, K_n$ to the end of the ordering.

Step 4. Perform a sequence of transpositions and ordered handle slides on the L_i's as before.

Step 5. Move the K_i's (via transpositions) back to the beginning of the ordering.

Step 6. Perform appropriate handle slides of the L_i's over the K_i's, yielding (L', s') plus n unknots with framing 1 (i.e. $K_1, \ldots, K_n$).

Step 7. Remove $K_1, \ldots, K_n$.

We must choose $b_1, \ldots, b_n$ so that the transpositions in steps 3, 4 and 5 are permissible. By (5.17) and (5.19) it suffices to find, for each OFL occuring in steps 3–5 and each $1 \leq k \leq 2n$, a choice of the b_i's such that $\det_k$ of that OFL is nonzero. In steps 3 and 5, $b_1 = \cdots = b_n = 0$ is such a choice. In step 4 with $k \leq n$, choose the b_i's large enough that the linking matrix restricted to the L_i's is positive definite. In step 4 with $k \geq n+1$, choose $b_1 = \cdots = b_n = 0$. $\qquad\square$

§ 6

Consequences of the Dehn Surgery Formula

In this section we use the Dehn surgery formula (4.2) to prove various things about λ.

(6.1) Proposition. *Let M_1 and M_2 be $\mathbf{Q}$HSs. Let $M_1 \# M_2$ denote the connected sum of M_1 and M_2. Then*

$$\lambda(M_1 \# M_2) = \lambda(M_1) + \lambda(M_2).$$

Proof: Let L_i be a permissible ordered framed link representing M_i, $i = 1, 2$ (see Section 5). Let $L_1 \cup L_2$ denote the permissible ordered framed link obtained by situating L_1 and L_2 in disjoint 3-balls in S^3 and putting L_2 after L_1 in the ordering. Then $L_1 \cup L_2$ represents $M_1 \# M_2$. The presence of the L_1 part of $L_1 \cup L_2$ does not affect homological information or Alexander polynomials of the L_2 part, so the proposition now follows easily from (4.2). $\square$

The following is a special case of (4.2).

(6.2) Proposition. *Let K be a null-homologous knot in a $\mathbf{Q}$HS M. Let $K_{p/q}$ denote p/q-surgery on K. Let $L_{p/q}$ denote the p, q-lens space (i.e. p/q-surgery on the unknot in S^3). Then*

$$\lambda(K_{p/q}) = \lambda(M) + \lambda(L_{p/q}) + \frac{q}{p}\Gamma(K).$$

$\square$

(4.1) and (4.2) imply

(6.3) **Proposition.** *Let $L_{p/q}$ denote the p,q-lens space. Then*

$$\lambda(L_{p/q}) = -s(q,p).$$

□

It is perhaps worth noting that (6.3) implies that $4p\lambda(L_{p/q})$ is equal to the signature defect of $L_{p/q}$ (see [HZ]).

(6.4) **Proposition.** $6|H_1(M;\mathbf{Z})|\lambda(M) = 6\bar{\lambda}(M) \in \mathbf{Z}$ *for all* **Q***HSs M.*

Proof: This can, in principle, be proved using (4.2), but it will be easier to use (3.34) and (3.36) (which are key ingredients in proving (4.2)).

By (5.4) and the fact that (6.4) is true for S^3, it suffices to show that $6\bar{\lambda}(K_a) - 6\bar{\lambda}(K_b) \in \mathbf{Z}$ for any knot K in a **Q**HS and any appropriate homology classes $a, b \in H_1(\partial(\mathrm{nbd}K);\mathbf{Z})$. This follows from (3.34), (3.36), the fact that $\bar{\lambda}'(K) \in \mathbf{Z}$ (which is clear from the proof of (3.34)), and the fact that $r(K, \cdot, \cdot) = 0$ (which was proved in Section 4).   □

The next result relates λ to the μ-invariant of a $\mathbf{Z}_2$-homology sphere ($\mathbf{Z}_2$HS). All facts concerning μ-invariants which are stated without proof can be found in [K2], or can be proved easily using the techniques of [K2] and will be proved in [W2].

Let M be an oriented 3-manifold and let σ be a spin structure on M. Let W be an oriented 4-manifold with spin structure ω such that $\partial W = M$ and ω restricts to σ. Define

$$\mu(M,\sigma) = \mathrm{sign}\,(W) \quad \mathrm{mod}\ 16,$$

where sign$\,(W)$ denotes the signature of the intersection pairing on $H_2(W;\mathbf{Z})$. This is well defined by Rochlin's theorem, which states that the signature of a closed spin 4-manifold is congruent to 0 mod 16. If M is a $\mathbf{Z}_2$HS then then it has a unique spin structure and we can speak unambiguously of $\mu(M)$.

(6.5) **Proposition.** *Let M be a* $\mathbf{Z}_2$*HS. Then*

$$4|H_1(M;\mathbf{Z})|^2\lambda(M) \equiv_{16} \mu(M)$$

(where $\equiv_{16}$ denotes congruence mod 16).

(Note: The case where M is an odd homology lens space has been proved independently in [BL].)

Proof: Define an *odd Dehn surgery* to be a Dehn surgery on a $\mathbf{Z}_2$HS which yields a $\mathbf{Z}_2$HS.

(6.6) **Lemma.** *Any $\mathbf{Z}_2 HS$ can be obtained from S^3 via a finite sequence of odd Dehn surgeries.*

Proof: Let M be a $\mathbf{Z}_2$HS. Represent M as surgery on an integral framed link $L = (L_1, \ldots, L_n)$ in S^3 (see Section 5). Let $A = [A_{ij}]$ be the linking matrix of L. Let d_k be the determinant of the $k \times k$ submatrix of A corresponding to the first k components of L. Then L represents a sequence of odd surgeries if and only if d_k is odd for $1 \leq k \leq n$. Thus it suffices to show that A can be transformed by row and column operations and stabilizations so that it has this property.

Let $\overline{A} = [\overline{A}_{ij}]$ be A reduced mod 2. After possibly stabilizing, we may assume that $\overline{A}$ has a 1 on the diagonal. Then arguing as in the proof of Theorem 4.3 of [MH], we see that $\overline{A}$ can be transformed via row and column operations into the identity matrix (over $\mathbf{Z}_2$). (The crucial ingredient in the argument is to see that any symmetric inner product space over $\mathbf{Z}_2$ has an element of square 0. But if $\langle a, a \rangle = \langle b, b \rangle = 1$ and $a \neq b$, then $\langle a + b, a + b \rangle = 0$.) Performing the corresponding operations on A produces a matrix with odd partial determinants. $\square$

In view of (6.6), it suffices to compute, modulo 16, how the left and right hand sides of (6.5) change when M is modified by an odd Dehn surgery, and to check that these changes are equal. We first consider the right hand side.

Let M be a $\mathbf{Z}_2$HS, let K be a knot in M, and let $N = M \setminus \mathrm{nbd}(K)$. Let $k \in H_1(\partial N; \mathbf{Z})$ be such that $l = dk$, where l is the longitude of N and $d \in \mathbf{Z}$. Then $H^1(N_k; \mathbf{Z}_2) \cong \mathbf{Z}_2$, so N_k has two spin structures, σ_1 and σ_2. $\mu(N_k, \sigma_1)$ differs from $\mu(N_k, \sigma_2)$ by a multiple of 8. Define the *arf invariant* of N to be

$$\mathrm{arf}(N) \stackrel{\mathrm{def}}{=} \frac{\mu(N_k, \sigma_1) - \mu(N_k, \sigma_2)}{8} \in \mathbf{Z}_2.$$

Let $m \in H_1(\partial N; \mathbf{Z})$ be such that $\langle m, k \rangle = 1$. Let $N_{p,q} \stackrel{\mathrm{def}}{=} N_{pm+qk}$. Let $L_{p/q}$ denote, as usual, the p, q lens space.

(6.7) **Lemma.** *For all relatively prime p, q such that $N_{p,q}$ is a $\mathbf{Z}_2 HS$ (i.e. such that p is odd),*

$$\mu(N_{p,q}) \equiv_{16} \mu(N_{1,0}) + \mu(L_{p/q}) + 8q \, \mathrm{arf}(N).$$

□

The case where M is a $\mathbf{Z}$HS is proved in [Go]. The proof of the general case is similar and will be given in [W2].

(6.8) Lemma.

$$\text{arf}(N) \equiv_2 \frac{1}{2} |\text{Tor}(H_1(N;\mathbf{Z}))|^2 \left(\Gamma(N) - \frac{d^2-1}{12} \right),$$

where $|\text{Tor}(H_1(N;\mathbf{Z}))|$ denotes the order of the the torsion subgroup of $H_1(N;\mathbf{Z})$. □

Again, this is well known for the case that M is a $\mathbf{Z}$HS, and the general case will be proved in [W2].

The next lemma is equivalent to (6.5) in the case where M is an odd lens space.

(6.9) Lemma. *Let p and q be relatively prime with p odd. Then*

$$\mu(L_{p/q}) \equiv_{16} -4p^2 s(q,p).$$

□

By the remark following (6.3), proving (6.9) amounts to establishing a relation between the μ-invariant of an odd lens space and its signature defect. As these two invariants are both defined in terms of bounded 4-manifolds, it is not surprising the such a relation exists. Details can be found in [BL], Lemma 4.4.

(6.10) Lemma. *If M is a $\mathbf{Z}_2$HS, then $\mu(M)$ is even.*

Proof: This follows from (6.6), (6.7), (6.9), and the fact that $-4p^2 s(q,p)$ is even if p is odd. □

(6.11) Lemma. *Let N and d be as above. Then*

$$\frac{2|\text{Tor}(H_1(N;\mathbf{Z}))|(d^2-1)}{3} \equiv_{16} 0.$$

Proof: If d is not divisible by 3, then $d^2 - 1$ is divisible by 24. If d is divisible by 3, then $d^2 - 1$ is divisible by 8 and $|\text{Tor}(H_1(N;\mathbf{Z}))|$ is divisible by 3. (The longitude represents an element of order d in $\text{Tor}(H_1(N;\mathbf{Z}))$.) □

Finally, note that

(6.12) **Lemma.** *Let $e, o \in \mathbf{Z}$, e even, o odd. Then*

$$o^2 e \equiv_{16} e.$$

$\square$

We are now ready to prove (6.5). Let N be as above. By (6.6), it suffices to show that (6.5) holds for $N_{p,q}$ (p odd, $p > 0$) assuming that it holds for, say, $N_{1,0}$. Let $h = |H_1(N_{1,0}; \mathbf{Z})|$. Note that $|H_1(N_{p,q}; \mathbf{Z})| = ph$ and $|\mathrm{Tor}(H_1(N; \mathbf{Z}))| = h/d$. We have

$$
\begin{aligned}
\mu(N_{p,q}) \;\equiv_{16}\;\; & \mu(N_{1,0}) + \mu(L_{p/q}) + 8q\,\mathrm{arf}(N) \\
& \text{(by (6.7))} \\[1em]
\equiv_{16}\;\; & 4p^2 h^2 \lambda(N_{1,0}) - 4p^2 h^2 s(q,p) + 4pq \left(\frac{h}{d}\right)^2 \left(\Gamma(N) - \frac{d^2-1}{12}\right) \\
& \text{(by the inductive assumption, (6.12), (6.9) and (6.8))} \\[1em]
=\;\; & 4p^2 h^2 \left(\lambda(N_{1,0}) + \tau(m, pm + qk, dk) + \right. \\[0.5em]
& \left. \frac{\langle m, pm + qk \rangle}{\langle m, dk \rangle \langle pm + qk, dk \rangle} \Gamma(N) \right) - 8qp \left(\frac{h^2}{d^2}\right) \frac{d^2-1}{12} \\
& \text{(by (4.1))} \\[1em]
\equiv_{16}\;\; & 4|H_1(N_{p,q}; \mathbf{Z})|^2 \lambda(N_{p,q}) \\
& \text{(by (4.2) and (6.11)).}
\end{aligned}
$$

This completes the proof of (6.5). $\square$

§ A

Dedekind Sums

In this section we prove various properties of the function τ. We give an alternative definition of τ ((A.4), below), show that this is equivalent to (4.1), and use the alternative definition to derive the properties of τ used in Sections 4 and 5. (Thus it is not necessary to ever mention Dedekind sums.) At the end, we show that the two definitions agree. The alternative definition is inspired by Theorem 1 of [H].

The treatment in this section is somewhat crude, but it has the advantage of being self-contained.

The alternative definition will require a few preliminary definitions.

Let $\widetilde{SL}_2(\mathbf{R})$ be the universal cover of $SL_2(\mathbf{R})$. $\widetilde{SL}_2(\mathbf{R})$ acts on $\widetilde{\mathbf{R}_*^2}$, the universal cover of $\mathbf{R}_*^2 \overset{\text{def}}{=} \mathbf{R}^2\backslash\{0\}$. This action commutes with the action of $\mathbf{R}_+$ on $\widetilde{\mathbf{R}_*^2}$, so there is an induced action of $\widetilde{SL}_2(\mathbf{R})$ on $\mathbf{R}_P \overset{\text{def}}{=} \widetilde{\mathbf{R}_*^2}/\mathbf{R}_+ \cong \mathbf{R}$. $\widetilde{SL}_2(\mathbf{R})$ acts on $\mathbf{R}_P$ via periodic diffeomorphisms, that is, diffeomorphisms which commute with the deck transformations of the covering

$$\mathbf{R}_P \to (\mathbf{R}_*^2)/\mathbf{R}_+ = S^1,$$

and we parameterize $\mathbf{R}_P$ so that a generator of the deck transformations is $x \mapsto x + 1$. (Note that $\widetilde{SL}_2(\mathbf{R})$ acts via diffeomorphisms of period $1/2$.) Give $\mathbf{R}^2$ its standard orientation, orient $\mathbf{R}_*^2$ and $\widetilde{\mathbf{R}_*^2}$ compatibly, and orient $\mathbf{R}_P$ so that its orientation preceded by the standard one on $\mathbf{R}_+$ gives the orientation on $\widetilde{\mathbf{R}^2\backslash\{0\}}$.

Let G be the group of diffeomorphisms of $\mathbf{R}$ with period 1. For $g \in G$, define

$$\text{rot}\,(g) \overset{\text{def}}{=} \lim_{n\to\infty} \frac{g^n(x) - x}{n}$$

where $x \in \mathbf{R}$. By a well known folk theorem, this limit exists and is independent of x. Furthermore,

$$\text{(A.1)} \qquad\qquad \text{rot}\,(hgh^{-1}) = \text{rot}\,(g)$$

for all $g, h \in G$. Since we have an action of $\widetilde{SL}_2(\mathbf{R})$ on $\mathbf{R}_P$ and an identification of $\mathbf{R}_P$ with $\mathbf{R}$ up to periodic diffeomorphisms, we get a map

$$\text{rot}\,:\widetilde{SL}_2(\mathbf{R}) \to \mathbf{R}.$$

Let $\widetilde{SL}_2(\mathbf{Z}) = \pi^{-1}(SL_2(\mathbf{Z})) \subset \widetilde{SL}_2(\mathbf{R})$. There is an exact sequence

$$1 \to K \to \widetilde{SL}_2(\mathbf{Z}) \to SL_2(\mathbf{Z}) \to 1,$$

where $K \cong \mathbf{Z}$. It is easy to verify that rot restricts to an isomorphism from K to $\mathbf{Z}$.

For H any Group, let H_{ab} denote $H/[H, H]$. Let $\varphi : \widetilde{SL}_2(\mathbf{Z}) \to \widetilde{SL}_2(\mathbf{Z})_{\text{ab}} \cong \mathbf{Z}$ be the projection. Clearly

$$\text{(A.2)} \qquad\qquad \varphi(hgh^{-1}) = \varphi(g)$$

Since $SL_2(\mathbf{Z})_{\text{ab}} \cong \mathbf{Z}_{12}$, $\varphi(K)$ has index 12 in $\widetilde{SL}_2(\mathbf{Z})_{\text{ab}}$. Fix an identification of $\widetilde{SL}_2(\mathbf{Z})_{\text{ab}}$ with $\mathbf{Z}$ by requiring that

$$\text{(A.3)} \qquad\qquad \varphi(k) = 12\,\text{rot}\,(k)$$

for all $k \in K$.

More concretely, let

$$U = \begin{pmatrix} 1 & 1 \\ 0 & 1 \end{pmatrix} \in SL_2(\mathbf{Z})$$

$$L = \begin{pmatrix} 1 & 0 \\ 1 & 1 \end{pmatrix} \in SL_2(\mathbf{Z}).$$

There are elements $\widetilde{U}$ and $\widetilde{L}$ of $\widetilde{SL}_2(\mathbf{Z})$ such that $\pi(\widetilde{U}) = U$, $\pi(\widetilde{L}) = L$, and such that $\widetilde{SL}_2(\mathbf{Z})$ is generated by $\widetilde{U}$ and $\widetilde{L}$ subject to the single relation

$$\widetilde{L}\widetilde{U}^{-1}\widetilde{L}\widetilde{U}\widetilde{L}^{-1}\widetilde{U} = 1.$$

In terms of these generators,

$$\varphi(\widetilde{U}) = -1$$
$$\varphi(\widetilde{L}) = 1.$$

Let V be a two dimensional integral lattice equiped with an antisymmetric inner product $\langle \cdot, \cdot \rangle$ such that $\langle a, b \rangle = \pm 1$ if and only if a, b is a basis of V. (For example, $V = H_1(T^2; \mathbf{Z})$ with the intersection pairing.) Choose an identification of V with $\mathbf{Z}^2$ (equiped with its standard pairing $\langle (a_1, a_2), (b_1, b_2) \rangle = a_1 b_2 - a_2 b_1$). This induces an identification of $\widehat{\mathrm{Aut}}\,(V)$ with $\widetilde{SL_2}(\mathbf{Z})$, and hence functions $\mathrm{rot} : \widehat{\mathrm{Aut}}\,(V) \to \mathbf{R}$ and $\varphi : \widehat{\mathrm{Aut}}\,(V) \to \mathbf{Z}$. It follows from (A.1) and (A.2) that these functions are independent of the choice of identification of V with $\mathbf{Z}^2$.

For $a, b, l \in V$ define

$$\delta(a, b; l) \stackrel{\text{def}}{=} \begin{cases} 0 & \text{if } \langle a, l \rangle \langle b, l \rangle > 0 \\ 1 & \text{otherwise} \end{cases}$$

and

$$\epsilon(a, b; l) \stackrel{\text{def}}{=} \begin{cases} 0 & \text{if } \langle a, b \rangle > 0 \\ 1 & \text{if } \langle a, b \rangle < 0 \text{ and } \delta(a, b; l) = 0 \\ -1 & \text{if } \langle a, b \rangle < 0 \text{ and } \delta(a, b; l) = 1 \\ 0 & \text{if } \langle a, b \rangle = 0 \text{ and } \delta(a, b; l) = 0 \\ -1 & \text{if } \langle a, b \rangle = 0 \text{ and } \delta(a, b; l) = 1. \end{cases}$$

Let V_{pr} denote the primitive elements of V, i.e. those that are not non-trivial multiples of other elements. Let $a, b \in V_{\mathrm{pr}}$ and $l \in V$. Choose $\hat{a}, \hat{b} \in V_{\mathrm{pr}}$ and $\hat{l} \in V \otimes \mathbf{Q}$ so that $\langle a, \hat{a} \rangle = \langle b, \hat{b} \rangle = \langle l, \hat{l} \rangle = 1$. Define $X \in \mathrm{Aut}\,(V)$ by

$$\begin{aligned} X(a) &= b \\ X(\hat{a}) &= \hat{b}. \end{aligned}$$

Choose $\tilde{X} \in \widehat{\mathrm{Aut}}\,(V)$ such that $\pi(\tilde{X}) = X$. Finally, define

$$(\text{A.4}) \quad \tau(a, b; l) = \frac{1}{12}\left(\frac{\langle a, \hat{l} \rangle}{\langle a, l \rangle} - \frac{\langle \hat{a}, l \rangle}{\langle a, l \rangle} - \frac{\langle b, \hat{l} \rangle}{\langle b, l \rangle} + \frac{\langle \hat{b}, l \rangle}{\langle b, l \rangle} \right)$$
$$+ \frac{1}{12}\varphi(\tilde{X}) - \left[\mathrm{rot}\,(\tilde{X}) + \frac{1}{2}\epsilon(a, b; l) \right] - \frac{1}{2}\delta(a, b; l).$$

($[x]$ denotes that greatest integer $\leq x$.)
(Note: Steve Boyer has pointed out to me that

$$\frac{1}{12}\varphi(\tilde{X}) - \left[\mathrm{rot}\,(\tilde{X}) + \frac{1}{2}\epsilon(a, b; l) \right] - \frac{1}{2}\delta(a, b; l) =$$
$$-\frac{1}{12}\Phi\left(\begin{bmatrix} \langle b, \hat{a} \rangle & \langle \hat{b}, \hat{a} \rangle \\ -\langle b, a \rangle & -\langle \hat{b}, a \rangle \end{bmatrix} \right) + \frac{1}{4}\mathrm{sign}\left(\frac{\langle a, b \rangle}{\langle a, l \rangle \langle b, l \rangle} \right),$$

where Φ is Radamacher's Φ-function

$$\Phi : SL_2(\mathbf{Z}) \;\to\; \mathbf{Z}$$

$$\Phi\left(\begin{bmatrix} w & x \\ y & z \end{bmatrix}\right) \;=\; \begin{cases} \frac{x}{z}, & y = 0 \\ -12s(z,y) + \frac{w+z}{y}, & y \neq 0 \end{cases}$$

(see [RG]).)

Before showing that τ is well defined and proving that it has the desired properties, we need the following six lemmas. The proofs are elementary and are left to the reader.

(A.5) **Lemma.** *Let $\tilde{X} \in \widetilde{SL}_2(\mathbf{R})$. Then $\pi(\tilde{X}) \in SL_2(\mathbf{R})$ has a positive [negative] eigenvalue if and only if* $\mathrm{rot}\,(\tilde{X}) \in \mathbf{Z}$ *[$\mathrm{rot}\,(\tilde{X}) \in \mathbf{Z} + \frac{1}{2}$].* $\square$

(A.6) **Lemma.** *Let $\tilde{X} \in \widetilde{SL}_2(\mathbf{R})$. Then $\mathrm{tr}(\pi(\tilde{X})) \geq 2$ [$\mathrm{tr}(\pi(\tilde{X})) \leq -2$] if and only if $\pi(\tilde{X})$ has a positive [negative] eigenvalue.* $\square$

(A.7) **Lemma.** *Let $\tilde{X}, \tilde{Y} \in \widetilde{SL}_2(\mathbf{Z})$. Then*

$$|\mathrm{rot}\,(\tilde{X}) + \mathrm{rot}\,(\tilde{Y}) - \mathrm{rot}\,(\tilde{X}\tilde{Y})| < 1.$$

$\square$

(For (A.7), keep in mind that elements of $\widetilde{SL}_2(\mathbf{Z})$ act on $\mathbf{R}_P$ with period $1/2$, not merely 1.)

(A.8) **Lemma.** *Let $\tilde{X}, \tilde{Y} \in \widetilde{SL}_2(\mathbf{Z})$. If $[\tilde{X}, \tilde{Y}] = 1$, then*

$$\mathrm{rot}\,(\tilde{X}) + \mathrm{rot}\,(\tilde{Y}) - \mathrm{rot}\,(\tilde{X}\tilde{Y}) = 0.$$

$\square$

For $a \in \mathbf{R}^2$ and $t \in \mathbf{R}$, define $S(a,t) \in SL_2(\mathbf{R})$ by $S(a,t)(b) = b + t\langle a, b\rangle a$. The mapping $t \mapsto S(a,t)$ is a homomorphism. Let $\tilde{S}(a,t)$ denote the lift to $\widetilde{SL}_2(\mathbf{R})$. If $a \in \mathbf{Z}^2$ and $t \in \mathbf{Z}$, then $S(a,t) \in SL_2(\mathbf{Z})$ and $\tilde{S}(a,t) \in \widetilde{SL}_2(\mathbf{Z})$. Let e_1 and e_2 be the standard basis of $\mathbf{Z}^2$. Note that $\tilde{U} = \tilde{S}(e_1, 1)$ and $\tilde{L} = \tilde{S}(e_2, -1)$, where $\tilde{U}$ and $\tilde{L}$ are as defined above.

(A.9) **Lemma.** *Let $\tilde{X} \in \widetilde{SL}_2(\mathbf{R})$ and $a \in \mathbf{R}^2$. Then there is an integer n such that*

$$\mathrm{rot}\,(\tilde{X}\tilde{S}(a,t)) \subset n + (\mathrm{sign}\,\langle a, \pi(\tilde{X})(a)\rangle)\,[0, \tfrac{1}{2}]$$

for all $t \in \mathbf{R}$. $\square$

(A.10) Lemma. *Let* $a \in \mathbf{Z}_{pr}^2$, $n \in \mathbf{Z}$. *Then*

$$\varphi(\widetilde{S}(a,n)) = -n.$$

□

We now show that $\tau(a,b;l)$ is independent of the choices of $\widetilde{X}$, $\hat{a}$, $\hat{b}$ and $\hat{l}$. Denote the terms on the right hand side of (A.4) by 1 ($\frac{1}{12}\frac{\langle a,l\rangle}{\langle a,l\rangle}$) through 7 ($-\frac{1}{2}\delta(a,b;l)$).

If we replace $\widetilde{X}$ with $k\widetilde{X}$, $k \in K$, then term 5 changes by rot(k) (by (A.3)), term 6 changes by $-$rot(k) (by (A.8) and the fact that K is contained in the center of $\widetilde{SL}_2(\mathbf{Z})$), and the other terms are not affected. Thus $\tau(a,b;l)$ does not depend on the choice of $\widetilde{X}$.

If we replace $\hat{a}$ with $\hat{a}+na$ ($n \in \mathbf{Z}$), we must replace X with $XS(a,-n)$ and $\widetilde{X}$ with $\widetilde{X}\widetilde{S}(a,-n)$. Terms 1, 3, 4 and 7 are not affected. By (A.9), term 6 does not change, while term 2 changes by $-\frac{n}{12}$ and (by (A.10)) term 5 changes by $\frac{n}{12}$. Thus $\tau(a,b;l)$ does not depend on the choice of $\hat{a}$.

Similar arguments show that $\tau(a,b;l)$ does not depend on the choices of $\hat{b}$ and $\hat{l}$. Therefore $\tau(a,b;l)$ is well-defined.

Now we show that τ has the properties required in sections 4 and 5. In the statements of the following lemmas, all variables (e.g. a, b, l) should be considered to be quantified by "for all", subject to the constraint that they lie in the domain of τ.

(A.11) Lemma. $\tau(-a,b;l) = \tau(a,-b;l) = \tau(a,b;-l) = \tau(a,b;l)$.

Proof: Changing the sign of a, b or l does not affect terms 1–4. Changing the sign of l does not affect terms 5–7.

There is a unique $\widetilde{R} \in \widetilde{SL}_2(\mathbf{Z})$ such that $\pi(\widetilde{R}) = \begin{bmatrix} -1 & 0 \\ 0 & -1 \end{bmatrix}$, rot$(\widetilde{R}) = \frac{1}{2}$ and $\varphi(\widetilde{R}) = 6$. Furthermore, $\widetilde{R}$ is contained in the center of $\widetilde{SL}_2(\mathbf{Z})$. If we change the sign of a or b, we must replace $\widetilde{X}$ with $\widetilde{R}\widetilde{X}$. Suppose $\delta(a,b;l) = 0$. Then $\delta(a,b;l)$ changes by 1, $\epsilon(a,b;l)$ changes by -1, and rot$(\widetilde{R}\widetilde{X}) = $ rot$(\widetilde{R}) + $ rot$(\widetilde{X}) = $ rot$(\widetilde{X}) + \frac{1}{2}$ (by (A.8)). Therefore term 5 changes by $\frac{1}{2}$, term 6 does not change, and term 7 changes by $-\frac{1}{2}$. The case $\delta(a,b;l) = 1$ is similar. □

(A.12) Lemma. τ *changes sign if the orientation on V changes.*

Proof: If we change the orientation on V, we must also change the signs of $\hat{a}$, $\hat{b}$ and $\hat{l}$. Thus terms 1–4 change sign. $\tilde{X}$ and $\delta(a,b;l)$ remain the same, but rot and φ (and hence term 5) change sign. After possibly changing the sign of (say) a, we may assume (by (A.11)) that $\delta(a,b;l) = 0$. By choosing $\hat{a}$ and $\hat{b}$ properly, we may assume that $\mathrm{rot}(\tilde{X}) \in \mathbf{Z}$. Hence term 6 and (trivially) term 7 also change sign. $\square$

(A.13) Lemma. $\tau(b,a;l) = -\tau(a,b;l)$.

Proof: This is similar to 3 above. $\square$

(A.14) Lemma. $\tau(a,b;l) + \tau(b,c;l) + \tau(c,a;l) = 0$.

Proof: After possibly changing the signs of a, b or c, we may assume (by (A.11)) that $\langle a,l \rangle$, $\langle b,l \rangle$, $\langle c,l \rangle > 0$. After possibly reordering a, b and c, we may assume (by (A.13)) that $\langle a,b \rangle$, $\langle b,c \rangle > 0$, $\langle c,a \rangle < 0$. (If, say, $\langle a,b \rangle = 0$ then we are done, by (A.13) and (A.11).) We can choose $\widetilde{X_{ab}}$, $\widetilde{X_{bc}}$ and $\widetilde{X_{ca}}$ so that $\widetilde{X_{ab}}\widetilde{X_{bc}}\widetilde{X_{ca}} = 1$. By choosing $\hat{a}$, $\hat{b}$ and $\hat{c}$ properly, we may assume that $\mathrm{rot}(\widetilde{X_{ab}})$, $\mathrm{rot}(\widetilde{X_{bc}})$ and $\mathrm{rot}(\widetilde{X_{ca}})$ are integers. By (A.7), this implies that $\mathrm{rot}(\widetilde{X_{ab}}) + \mathrm{rot}(\widetilde{X_{bc}}) + \mathrm{rot}(\widetilde{X_{ca}}) = 0$.

The contributions of terms 1–4 to $\tau(a,b;l) + \tau(b,c;l) + \tau(c,a;l)$ clearly cancel. The contribution of term 5 cancels because φ is a homomorphism. Since $\epsilon(a,b;l)$, $\epsilon(b,c;l)$ and $\epsilon(c,a;l)$ are either 0 or 1, the contribution from term 6 cancels. The contribution from term 7 cancels because $\delta(a,b;l) = \delta(b,c;l) = \delta(c,a;l) = 0$. $\square$

(A.15) Lemma. $\tau(a + n\langle l,a \rangle l, b + m\langle l,b \rangle l; l) = \tau(a,b;l)$ for all $m, n \in \mathbf{Z}$ if $l \in V_{pr}$.

Proof: Suppose we replace a with $a + n\langle l,a \rangle l = S(l,n)(a)$. Then we must also replace $\hat{a}$ with $\hat{a} + n\langle l,\hat{a} \rangle l$ and $\tilde{X}$ with $\tilde{X}\tilde{S}(l,-n)$. Thus term 1 changes by $-\frac{n}{12}$, term 5 changes by $\frac{n}{12}$ (by (A.10)), and terms 2, 3, 4 and 7 remain the same.

Suppose that $n < 0$. After possibly changing the sign of a and the orientation on V, we may assume (by (A.11) and (A.12)) that $\delta(a,b;l) = 0$ and $\langle a,b \rangle > 0$ or $a = b$. Under these circumstances the proper choice of $\hat{a}$ and $\hat{b}$ will guarantee that

$$\begin{aligned} \mathrm{tr}(\pi(\tilde{X}\tilde{S}(l,t))) \;&=\; -\langle \hat{a} + t\langle l,\hat{a} \rangle l, b \rangle + \langle a + t\langle l,a \rangle l, \hat{b} \rangle \\ &\geq\; 2 \end{aligned}$$

for all real $0 \leq t \leq -n$. Therefore (by (A.5) and (A.6)) $\mathrm{rot}(\tilde{X}\tilde{S}(l,-n)) = \mathrm{rot}(\tilde{X}) \in \mathbf{Z}$. Since $\epsilon(a,b;l) = 0$ or 1 and $\epsilon(a + n\langle l,a \rangle l, b; l) = 0$ or 1 (because

$\delta(a,b;l) = 0$ and $\delta(a + n\langle l, a\rangle l, b; l) = 0)$, term 6 does not change. The case $n > 0$ is similar. $\qquad\square$

(A.16) **Lemma.** $\tau(a, b; l_1) - \tau(a, b; l_2) =$

$$-\frac{1}{12}\left(\frac{\langle a, b\rangle}{\langle a, l_1\rangle\langle b, l_1\rangle} - \frac{\langle a, b\rangle}{\langle a, l_2\rangle\langle b, l_2\rangle} - \frac{\langle l_1, l_2\rangle}{\langle l_1, a\rangle\langle l_2, a\rangle} + \frac{\langle l_1, l_2\rangle}{\langle l_1, b\rangle\langle l_2, b\rangle}\right)$$
$$+ \frac{1}{4}\left(\text{sign}\,\frac{\langle a, b\rangle}{\langle a, l_1\rangle\langle b, l_1\rangle} - \text{sign}\,\frac{\langle a, b\rangle}{\langle a, l_2\rangle\langle b, l_2\rangle}\right).$$

Proof: Since both sides of the identity asserted in (A.16) are antisymmetric with respect to $a \leftrightarrow b$, it suffices to consider the case $\langle a, b\rangle > 0$. (The case $\langle a, b\rangle = 0$ follows from (A.11) and (A.13).) The contribution of terms 1–4 to $\tau(a, b; l_1) - \tau(a, b; l_2)$ is

$$\frac{1}{12}\left(\frac{\langle a, \hat{l_1}\rangle\langle b, l_1\rangle - \langle \hat{a}, l_1\rangle\langle b, l_1\rangle - \langle b, \hat{l_1}\rangle\langle a, l_1\rangle + \langle \hat{b}, l_1\rangle\langle a, l_1\rangle}{\langle a, l_1\rangle\langle b, l_1\rangle}\right.$$
$$\left. - \frac{\langle a, \hat{l_2}\rangle\langle b, l_2\rangle - \langle \hat{a}, l_2\rangle\langle b, l_2\rangle - \langle b, \hat{l_2}\rangle\langle a, l_2\rangle + \langle \hat{b}, l_2\rangle\langle a, l_2\rangle}{\langle a, l_2\rangle\langle b, l_2\rangle}\right).$$

Upon repeated applications of the identity

$$\langle x, y\rangle = \langle x, \hat{z}\rangle\langle z, y\rangle - \langle x, z\rangle\langle \hat{z}, y\rangle$$

(where $\langle z, \hat{z}\rangle = 1$), this becomes

$$-\frac{1}{12}\left(\frac{\langle a, b\rangle}{\langle a, l_1\rangle\langle b, l_1\rangle} - \frac{\langle a, b\rangle}{\langle a, l_2\rangle\langle b, l_2\rangle} - \frac{\langle l_1, l_2\rangle}{\langle l_1, a\rangle\langle l_2, a\rangle} + \frac{\langle l_1, l_2\rangle}{\langle l_1, b\rangle\langle l_2, b\rangle}\right).$$

The contributions from term 5 clearly cancel. Since $\langle a, b\rangle > 0$, $\epsilon(a, b; l_i) = 0$ and hence the contribution from term 6 is zero. Keeping in mind that $\langle a, b\rangle > 0$, we see that the contribution from term 7 is

$$-\frac{1}{2}\delta(a, b; l_1) + \frac{1}{2}\delta(a, b; l_2) = \frac{1}{4}\left(\text{sign}\,\frac{\langle a, b\rangle}{\langle a, l_1\rangle\langle b, l_1\rangle} - \text{sign}\,\frac{\langle a, b\rangle}{\langle a, l_2\rangle\langle b, l_2\rangle}\right)$$

$\qquad\square$

(A.17) **Lemma.** *Let a, b and l be such that $\langle a, b\rangle = 1$ and $\langle a, l\rangle\langle b, l\rangle > 0$.*
Then

$$\langle a, l\rangle\tau(a + b, a; l) + \langle b, l\rangle\tau(a + b, b; l) = \frac{1}{6}(\langle b, l\rangle - \langle a, l\rangle).$$

Proof: By (A.15) and (A.13)

$$\tau(a + b, a; b) = \tau(a, a; b) = 0.$$

Therefore, by (A.16),

$$\langle a, l\rangle\tau(a + b, a; l) = -\frac{\langle a, l\rangle}{12}\left(\frac{\langle b, a\rangle}{\langle a + b, l\rangle\langle a, l\rangle} - \frac{\langle b, a\rangle}{\langle a, b\rangle\langle a, b\rangle}\right.$$
$$\left. - \frac{\langle l, b\rangle}{\langle l, a + b\rangle\langle b, a + b\rangle} + \frac{\langle l, b\rangle}{\langle l, a\rangle\langle b, a\rangle}\right).$$

Similarly,

$$\langle b, l\rangle\tau(a + b, b; l) = -\frac{\langle b, l\rangle}{12}\left(\frac{\langle a, b\rangle}{\langle a + b, l\rangle\langle b, l\rangle} - \frac{\langle a, b\rangle}{\langle b, a\rangle\langle b, a\rangle}\right.$$
$$\left. - \frac{\langle l, a\rangle}{\langle l, a + b\rangle\langle a, a + b\rangle} + \frac{\langle l, a\rangle}{\langle l, b\rangle\langle a, b\rangle}\right).$$

Adding these two equations establishes the lemma.   □

It is perhaps worth noting that τ is uniquely determined by (A.14), (A.15) and (A.16).

Finally we show that the above definition of τ, (A.4), agrees with (4.1).

Choose an identification of V with $\mathbf{Z}^2$ so that $l = (d, 0)$ for some $d \in \mathbf{Z}$. Let $a = (u, v)$ and $b = (p, q)$. Without loss of generality (see (A.11)), $q, v > 0$.

First we calculate $\tau((0, 1), (p, q); (1, 0))$. Choose a continued fraction expansion $p/q = \langle a_0, a_1, \ldots, a_r\rangle$ of p/q. We may assume that $a_i \in \mathbf{Z}$, r is even, and $a_1, \ldots, a_r > 0$. (See [NZ] for these and other elementary facts about continued fractions.) Let

$$X = U^{a_0}L^{a_1}U^{a_2}\cdots L^{a_{r-1}}U^{a_r}.$$

Note that $X(0, 1) = (p, q)$. Let $(\hat{p}, \hat{q}) = X(-1, 0)$. Then $\langle(p, q), (\hat{p}, \hat{q})\rangle = 1$ and

$$\frac{\hat{p}}{\hat{q}} = -\langle a_0, \ldots, a_{r-1}\rangle.$$

Note also that

$$\frac{\hat{q}}{q} = -\langle 0, a_r, \ldots, a_1\rangle.$$

Let $\tilde{X} = \tilde{U}^{a_0}\tilde{L}^{a_1}\tilde{U}^{a_2}\dots\tilde{L}^{a_{r-1}}\tilde{U}^{a_r}$. It follows that

$$\varphi(\tilde{X}) = -a_0 + a_1 - a_2 + \dots + a_{r-1} - a_r.$$

Since $a_1,\dots,a_r > 0$, $\mathrm{rot}\,(\tilde{X}) \in [0, 1/2]$ and $\mathrm{rot}\,(\tilde{X}) = 0$ if $p \geq 0$. Since

$$\delta((0,1),(p,q);(1,0)) = 0,$$

$\epsilon((0,1),(p,q);(1,0)) = 0$ if $p \leq 0$ and $\epsilon((0,1),(p,q);(1,0)) = 1$ if $p > 0$.
Hence

$$\left[\mathrm{rot}\,(\tilde{X}) + \frac{1}{2}\epsilon((0,1),(p,q);(1,0))\right] = 0.$$

By Theorem 1 of [H],

$$s(p,q) \;=\; \frac{1}{12}\left(\langle a_0, a_1, \dots, a_r\rangle - \langle 0, a_r, \dots, a_1\rangle\right.$$
$$\left. - a_0 + a_1 - a_2 + \dots + a_{r-1} - a_r\right).$$

Putting the above facts together we have

$$\tau((0,1),(p,q);(1,0)) \;=\; \frac{1}{12}\left(\frac{0}{1} - \frac{0}{1} - \frac{p}{-q} + \frac{-\hat{q}}{q}\right) + \frac{1}{12}\varphi(\tilde{X})$$
$$= \;\frac{1}{12}\left(\langle a_0, a_1, \dots, a_r\rangle - \langle 0, a_r, \dots, a_1\rangle\right.$$
$$\left. -a_0 + a_1 - a_2 + \dots + a_{r-1} - a_r\right)$$
$$= \;s(p,q).$$

Similarly,

$$\tau((u,v)(0,1);(1,0)) = -s(u,v).$$

Therefore, by (A.14) and (A.16),

$$\tau((u,v),(p,q);(d,0)) = -s(u,v) + s(p,q) + \frac{(d^2-1)}{12}\frac{(uq - vp)}{(dv)(dq)}.$$

This is equivalent to (4.1).

§ B

Alexander Polynomials

This section contains facts about Alexander polynomials (and second derivatives evaluated at 1 thereof) needed elsewhere in the paper. First we define Alexander polynomials. (For more details, see [M1].)

Let R be a commutative ring and M be a finitely generated R-module. Let M have generators $x_1, \ldots, x_n$ subject to the relations $\sum a_{ij} x_j = 0$, $1 \leq i \leq m$. The ideal of R generated by all $n \times n$ determinants of the matrix $[a_{ij}]$ is called the *order ideal* on M and denoted $o(M)$. (If $m < n$, $o(M) \stackrel{\text{def}}{=} 0$.) This is independent of the choice of presentation of M. o is multiplicative in the following sense.

(B.1) **Proposition.** *If* $M_1 \subset M_2$ *are* R-modules, then

$$o(M_2) = o(M_1)o(M_2/M_1).$$

$\square$

If $o(M)$ is principal, let $\Delta(M)$ denote a generator of $o(M)$. $\Delta(M)$ is well-defined up to units in R.

If $\widetilde{X} \to X$ is a covering with group of deck transformations D, then $H_1(\widetilde{X}; \mathbf{Q})$ is a module over $\mathbf{Q}[D]$, and we define $o(\widetilde{X})$ to be $o(H_1(\widetilde{X}; \mathbf{Q}))$. If $D \cong \mathbf{Z}$, then $\mathbf{Q}[D]$ is a principal ideal domain, and we define $\Delta(\widetilde{X}) \in \mathbf{Q}[D]$ to be $\Delta(H_1(\widetilde{X}; \mathbf{Q}))$. This is well-defined up to units in $\mathbf{Q}[D]$, that is, elements of the form ct^k, where $c \in \mathbf{Q}$, t is a generator of D, and $k \in \mathbf{Z}$.

We are particularly interested in the case where X is the complement of a tubular neighborhood of a knot in a **QHS**, or, equivalently, X is a compact 3-manifold with $H_*(X; \mathbf{Q}) \cong H_*(S^1; \mathbf{Q})$. In this case the *universal* *free abelian cover* $\widetilde{X}_{FA}$ of X (determined by $\pi_1(X) \to \text{Free}\,(H_1(X;\mathbf{Z})) \cong \mathbf{Z}$) has deck transformations isomorphic to $\mathbf{Z}$ and we define $\Delta(X)$ (also denoted Δ_X), the *Alexander polynomial* of X, to be $\Delta(\widetilde{X}_{FA})$. Another case of interest is when X is the complement of a two component link in a **QHS**. $\widetilde{X}_{FA}$ has deck translations $\text{Free}\,(H_1(X;\mathbf{Z})) \cong \mathbf{Z} \oplus \mathbf{Z}$.

(B.2) Lemma. *Let X be the complement of a two component link in a* **QHS**. *Then $H_1(\widetilde{X}_{FA};\mathbf{Q})$ has a square presentation matrix over $\mathbf{Q}[\mathbf{Z} \oplus \mathbf{Z}]$.*

It follows that $o(\widetilde{X}_{FA})$ is principal, and we define $\Delta(X)$ to be $\Delta(\widetilde{X}_{FA})$.

Proof: Choose a basis x, y of $\text{Free}\,(H_1(X;\mathbf{Z})) \cong \mathbf{Z}^2$. There is a 2-complex $Y \subset X$ onto which X deformation retracts. For notational convenience, we now redefine X to be Y. Without loss of generality, X contains a single 0-cell p. Let $e_1, \ldots, e_n$ be the 1-cells of X and $f_1, \ldots, f_m$ be the 2-cells of X. Since the Euler characteristic of X is zero, $m = n - 1$. Without loss of generality

$$\begin{aligned} [e_1] &= x \in \text{Free}\,(H_1(X;\mathbf{Z})) \\ [e_2] &= y \in \text{Free}\,(H_1(X;\mathbf{Z})). \end{aligned}$$

Let $\tilde{p}$ be a fixed lift of p in $\widetilde{X}_{FA}$. Choose lifts $\tilde{e}_1, \ldots, \tilde{e}_n$ of $e_1, \ldots, e_n$ so that for all i and some $a_i, b_i \in \mathbf{Z}$

$$\partial(\tilde{e}_i) = (a_i x + b_i y)\tilde{p} - \tilde{p}.$$

(It follows that $a_1 = 1$, $b_1 = 0$, $a_2 = 0$, $b_2 = 1$.) Define the following 1-cycles in $\widetilde{X}_{FA}$:

$$l_1 \overset{\text{def}}{=} \tilde{e}_1 + (x)\tilde{e}_2 - (y)\tilde{e}_1 - \tilde{e}_2$$

$$l_{i-1} \overset{\text{def}}{=} \tilde{e}_i - \text{sign}\,(a_i)(\sum_{k=0}^{a_i} kx + b_i y)\tilde{e}_1 - \text{sign}\,(b_i)(\sum_{k=0}^{b_i} ky)\tilde{e}_2,$$

$3 \le i \le n$ (see Figure B.1).

Let $\widetilde{X}_{FA}^1$ denote the 1-skeleton of $\widetilde{X}_{FA}$. It is not hard to see that the classes $[l_1], \ldots, [l_{n-1}]$ freely generate $H_1(\widetilde{X}_{FA}^1;\mathbf{Q})$ (over $\mathbf{Q}[\text{Free}\,(H_1(X;\mathbf{Z}))]$). There is an exact sequence

$$\text{(B.3)} \qquad H_2(\widetilde{X}_{FA}, \widetilde{X}_{FA}^1;\mathbf{Q}) \overset{\partial}{\to} H_1(\widetilde{X}_{FA}^1;\mathbf{Q}) \to H_1(\widetilde{X}_{FA};\mathbf{Q}) \to 0.$$

$H_2(\widetilde{X}_{FA}, \widetilde{X}_{FA}^1;\mathbf{Q})$ is freely generated by $n-1$ lifts of the 2-cells $f_1, \ldots, l_{n-1}$ (one lift for each 2-cell). Hence (B.3) is equivalent to an $(n-1)$ by $(n-1)$

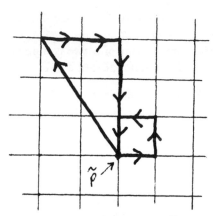

Figure B.1: Cycles in $\widetilde{X}_{FA}$.

presentation matrix for $H_1(\widetilde{X}_{FA}; \mathbf{Q})$. (Note that if X is the complement of a two component link of linking number zero, then this matrix will have rank $n - 2$.)   □

Let X be the complement of a two component link in a **QHS**. Let

$$\varphi : \mathrm{Free}\,(H_1(X;\mathbf{Z})) \to \mathbf{Z}$$

be a surjective homomorphism. Let t be a generator of $\mathbf{Z}$, with group composition written multiplicatively. Let $\varphi_* : \mathbf{Q}[\mathrm{Free}\,(H_1(X;\mathbf{Z}))] \to \mathbf{Q}[\mathbf{Z}] = \mathbf{Q}[t, t^{-1}]$ be the induced map. Let $\widetilde{X}$ be the cover of X defined by the homomorphism $\pi_1(X) \to \mathrm{Free}\,(H_1(X;\mathbf{Z})) \xrightarrow{\varphi} \mathbf{Z}$.

The next lemma is used in Section 5.

(B.4) Lemma. $\Delta(\widetilde{X}) = (1 - t)\varphi_*(\Delta(\widetilde{X}_{FA}))$.

Proof: Retain notation from the proof of (B.2). The 2-complex Y (alias X) can be chosen so that $\varphi([e_1]) = t$ and $[e_2]$ generates the kernel of φ. Let $\widetilde{X}^1$ denote the 1-skeleton of $\widetilde{X}$. Let $\pi : \widetilde{X}_{FA} \to \widetilde{X}$ be the natural projection, and let π also denote its restriction to 1-skeletons.

Let k be the 1-cycle in $\widetilde{X}$ consisting of the lift of e_2 which contains $\pi(\tilde{p})$. $H_1(\widetilde{X}; \mathbf{Q})$ is freely generated (over $\mathbf{Q}[t, t^{-1}]$) by $[k]$ and $[\pi(l_i)]$, $2 \leq i \leq n-1$. Note that

$$[\pi(l_1)] = (1 - t)[k].$$

Let f be a 2-cell of X and $\tilde{f}$ be a lift of f to $\tilde{X}_{FA}$ Let

$$[\partial(\tilde{f})] = \sum_{1}^{n-1} x_i[l_i],$$

where $x_i \in \mathbf{Q}[\text{Free}\,(H_1(X;\mathbf{Z}))]$. Then

$$[\partial(\pi(\tilde{f}))] = \sum_{1}^{n-1} \varphi_*(x_i)[\pi(l_i)].$$

It follows that a square presentation matrix for $H_1(\tilde{X};\mathbf{Q})$ is obtained by taking the square presentation matrix for $H_1(\tilde{X}_{FA};\mathbf{Q})$ of (B.2), applying φ_* to the entries, and multiplying the first column by $1 - t$. □

Let $\sum c_j[x_j] \in \mathbf{Q}[\mathbf{Z}]$. Let $\langle c \rangle \overset{\text{def}}{=} \sum c_j$ and $\langle cx \rangle \overset{\text{def}}{=} \sum c_j x_j$. If $\langle c \rangle \neq 0$, define

$$\Gamma(\sum c_j[x_j]) \overset{\text{def}}{=} \frac{\sum c_j \left(x_j - \frac{\langle cx \rangle}{\langle c \rangle} \right)^2}{\langle c \rangle}.$$

(If we think of $\sum c_j[x_j]$ as representing masses c_j at positions $x_j \in \mathbf{Z} \subset \mathbf{R}$, then Γ is the moment of inertia about the center of mass, divided by the total mass.) Note that Γ does not change if we multiply by units of $\mathbf{Q}[\mathbf{Z}]$. Thus $\Gamma(\Delta(M))$ is well-defined, so long as the sum of the coefficients of $\Delta(M)$ is not zero. If N is a knot complement (in a $\mathbf{Q}$HS), it is well known that $\Delta(N)$ has the form

$$\Delta(N) = \sum_{j} c_j (t^j + t^{-j}),$$

where t is a generator of $\mathbf{Z}$ written multiplicatively and j ranges over a finite set of half integers. (We are implicitly extending $\mathbf{Q}[\mathbf{Z}]$ to $\mathbf{Q}[\frac{1}{2}\mathbf{Z}]$. The case of half odd integers arises when the multiplicity of the longitude of the knot is even. See the proof of (B.10).) If we normalize so that $\Delta(N)|_{t=1} = 1$ (which is possible), then

(B.5) $\Gamma(N) \overset{\text{def}}{=} \Gamma(\Delta(N)) = 2 \sum_{j=0}^{n} c_j j^2 = \dfrac{d^2}{dt^2}\Delta(N)|_{t=1}.$

We now relate Γ to Seifert matrices. (What follows is standard material (see, for example, [Ro]), except for complications arising from the fact that the longitude is not necessarily a primitive element of $H_1(\partial N;\mathbf{Z})$.)

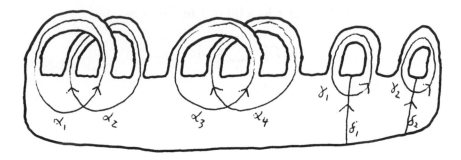

Figure B.2: Curves on E.

Let $E \subset N$ be a Seifert surface, that is, a connected, properly embedded surface representing a generator of $H_2(N, \partial N; \mathbf{Z}) \cong \mathbf{Z}$. Further assume that ∂E consists of d parallel (as opposed to antiparallel) curves on ∂N. (This can always be arranged.) Let $\alpha_1, \ldots, \alpha_{2g}, \gamma_1, \ldots, \gamma_{d-1}$ be simple closed curves representing a basis of $H_1(E; \mathbf{Q})$, as shown in Figure B.2. Orient the α_is so that

$$\langle \alpha_i, \alpha_j \rangle = \begin{cases} 1, & i \text{ odd}, j = i+1 \\ -1, & i \text{ even}, j = i-1 \\ 0, & \text{otherwise.} \end{cases}$$

Let Y be the complement of an open regular neighborhood of E in N. ∂Y contains two copies of E; call them E^+ and E^-. Let $\alpha_i^\pm \subset E^\pm$ be the curve corresponding to α_i. Let $a_i^\pm = [\alpha_i^\pm] \in H_1(Y; \mathbf{Q})$. Define $\gamma_i^\pm \subset E^\pm$ and $c_i^\pm \in H_1(Y; \mathbf{Q})$ similarly. $\widetilde{N}_{\mathrm{FA}}$ can be constructed by taking $\mathbf{Z}$ copies of Y and identifying E^+ of the ith copy with E^- of the $i+1^{\text{st}}$ copy. Identify Y with the zeroth copy of Y. Let $e_1, \ldots, e_n$ be a basis of $H_1(Y; \mathbf{Q})$. $H_1(\widetilde{N}_{\mathrm{FA}}; \mathbf{Q})$ is generated over $\mathbf{Q}[\mathbf{Z}] \cong \mathbf{Q}[t, t^{-1}]$ by $e_1, \ldots, e_n$ subject to the relations

(B.6) $\qquad \begin{cases} a_j^+ = ta_j^-, & 1 \le j \le 2g \\ c_j^+ = tc_j^-, & 1 \le j \le d-1 \end{cases}$

In order to get an explicit presentation matrix for $H_1(\widetilde{N}_{\mathrm{FA}}; \mathbf{Q})$, we need to be more explicit about the basis $e_1, \ldots, e_n$. Let M be a $\mathbf{Q}$HS which contains N as the complement of an open solid torus U. Let $E^* \overset{\text{def}}{=} E \cup U$. By Alexander duality, $H_1(Y; \mathbf{Q}) \cong H_1(E^*; \mathbf{Q})$. More concretely, let $\epsilon_1, \ldots, \epsilon_n$ be curves representing a basis of $H_1(E^*; \mathbf{Q})$. Let V be the $\mathbf{Q}$-vector space with basis $\epsilon_1, \ldots, \epsilon_n$. Let $\eta \subset Y$ be a curve. Then the correspondence

$$[\eta] \mapsto \sum_i \mathrm{lk}(\eta, \epsilon_i)\epsilon_i$$

gives rise to to an isomorphism from $H_1(Y; \mathbf{Q})$ to V. For the ϵ_i s we can take $\alpha_1, \ldots, \alpha_{2g}, \delta_1, \ldots, \delta_{d-1}$, where δ_i consists of an arc in E joining the 0th and i^{th} boundary components together with an arc in U (see Figure B.2). Orient the γ_is and δ_is so that

$$\langle \gamma_i, \delta_j \rangle = \begin{cases} 1, & i = j \\ 0, & \text{otherwise.} \end{cases}$$

With the above remarks in mind, it is easy to see that (B.6) gives rise to a presentation matrix

$$\begin{pmatrix} A & B \\ C & D \end{pmatrix},$$

where

$$\begin{aligned}
A_{ij} &= \operatorname{lk}(\alpha_i^+, \alpha_j) - t\operatorname{lk}(\alpha_i^-, \alpha_j) \\
B_{ij} &= \operatorname{lk}(\alpha_i^+, \delta_j) - t\operatorname{lk}(\alpha_i^-, \delta_j) \\
C_{ij} &= \operatorname{lk}(\gamma_i^+, \alpha_j) - t\operatorname{lk}(\gamma_i^-, \alpha_j) \\
D_{ij} &= \operatorname{lk}(\gamma_i^+, \delta_j) - t\operatorname{lk}(\gamma_i^-, \delta_j).
\end{aligned}$$

If $E^\pm$ is deformed slightly near its boundary, then $\partial E^\pm$ consists of d copies of $\gamma_i^\pm$. Since α_j is disjoint from $E^\pm$, $\operatorname{lk}(\gamma_i^\pm, \alpha_j) = \frac{1}{d}\langle \alpha_j, E^\pm \rangle = 0$, and hence $C = 0$. Therefore

$$(\text{B.7}) \qquad\qquad \Delta(N) = \det(A)\det(D).$$

If every entry of A and D is multiplied by $t^{-1/2}$, that is

$$\begin{aligned}
A_{ij} &= t^{-1/2}\operatorname{lk}(\alpha_i^+, \alpha_j) - t^{1/2}\operatorname{lk}(\alpha_i^-, \alpha_j) \\
D_{ij} &= t^{-1/2}\operatorname{lk}(\gamma_i^+, \delta_j) - t^{1/2}\operatorname{lk}(\gamma_i^-, \delta_j),
\end{aligned}$$

then (B.7) changes by a unit (of $\mathbf{Q}[t^{1/2}, t^{-1/2}]$) to become symmetric in $t^{1/2}$ and $t^{-1/2}$. Since this is a desirable property for $\Delta(N)$ to have, consider this change to have been effected in what follows.

If $\beta_1, \beta_2 \subset E$ are simple closed curves, then

$$\operatorname{lk}(\beta_1^+, \beta_2) - \operatorname{lk}(\beta_1^-, \beta_2) = \langle \beta_1, \beta_2 \rangle.$$

It follows that

$$\begin{aligned}
A_{ij}|_{t=1} &= \langle \alpha_i, \alpha_j \rangle \\
D_{ij}|_{t=1} &= \langle \gamma_i, \delta_j \rangle,
\end{aligned}$$

and hence that $\det(A)|_{t=1} = \det(D)|_{t=1} = 1$. Thus (B.7) is the symmetric, normal form of $\Delta(N)$, and

(B.8) $$\Gamma(N) = \left.\frac{d^2}{dt^2}\right|_{t=1} (\det(A)\det(D)).$$

Using the fact that

$$\left.\frac{d}{dt}\right|_{t=1} (\det(A)) = \left.\frac{d}{dt}\right|_{t=1} (\det(D)) = 0,$$

(B.8) becomes

(B.9) $$\Gamma(N) = \left.\frac{d^2}{dt^2}\right|_{t=1} (\det(A)) + \left.\frac{d^2}{dt^2}\right|_{t=1} (\det(D)).$$

It is not hard to verify that

$$D_{ij} = \begin{cases} \frac{i}{d}(t^{1/2} - t^{-1/2}), & j < i \\ \frac{i}{d}(t^{1/2} - t^{-1/2}) + t^{-1/2}, & j = i \\ (\frac{i}{d} - 1)(t^{1/2} - t^{-1/2}), & j > i. \end{cases}$$

A fun-filled exercise in algebraic manipulation (left to the reader, of course) now yields

(B.10)$\Gamma(N) = 2 \displaystyle\sum_{1 \le i,j \le g} \det \begin{bmatrix} \text{lk}(\alpha_{2i-1}^-, \alpha_{2j-1}) & \text{lk}(\alpha_{2i-1}^+, \alpha_{2j}) \\ \text{lk}(\alpha_{2i}^-, \alpha_{2j-1}) & \text{lk}(\alpha_{2i}^+, \alpha_{2j}) \end{bmatrix} + \dfrac{d^2 - 1}{12}.$

Bibliography

[AM]   S. Akbulut and J. McCarthy, Casson's invariant for oriented homology 3-spheres – an exposition, Princeton mathematical notes 36, Princeton university press, 1990.

[AS4]   M. F. Atiyah and I. M. Singer, The index of elliptic operators: IV, Ann. Math. **93** (1971), 119-138.

[AS]   M. F. Atiyah and I. M. Singer, Dirac operators coupled to vector potentials, Proc. Nat. Acad. Sci. **81** (1984), 2597-2600.

[BL]   S. Boyer and D. Lines, Surgery formulae for Casson's invariant and extensions to homology lens spaces, UQAM Rapport de recherche 66 (1988).

[BN]   S. Boyer and A. Nicas, Varieties of group representations and Casson's invariant for rational homology 3-spheres, preprint (1986).

[G1]   W. Goldman, The symplectic nature of the fundamental groups of surfaces, Adv. Math. **54** (1984), 200-225.

[G2]   W. Goldman, Representations of fundamental groups of surfaces, in Proceedings of Special Year in Topology, Maryland 1983-1984, Lecture notes in mathematics 1167, Springer-Verlag, 1985.

[G3]   W. Goldman, Invariant functions on Lie groups and hamiltonian flows on surface group representations, Invent. math. **85** (1986), 263-302.

[GM]   W. Goldman and J. Millson, The deformation theory of representations of fundamental groups of compact Kähler manifolds, Publ. Math. I.H.E.S. **67** (1988), 43-96.

[Go]   C. Gordon, Knots, homology spheres, and contractable 4-manifolds, Topology **14** (1975), 151-172.

[H]   D. Hickerson, Continued fractions and density results for Dedekind sums, J. Reine Angew. Math. **290** (1977), 113-116.

[HZ]   F. Hirzebruch and D. Zagier, The Atiyah-Singer index theorem and elementary number theory, Publish or perish, 1974.

[K1]   R. Kirby, A calculus for framed links in S^3, Invent. math. **45** (1978), 35-56.

[K2]   R. Kirby, The topology of 4-manifolds, Lecture notes in mathematics 1374, Springer-Verlag, 1989.

[L]   W. B. R. Lickorish, A representation of orientable combinatorial 3-manifolds, Ann. Math. **76** (1962), 531-538.

[M1]   J. W. Milnor, Infinite cyclic coverings, in Conference on the topology of manifolds (Michigan State Univ., East Lansing Mich., 1967), Prindle, Weber and Schmidt, 1968.

[M2]   J. W. Milnor, A duality theorem for Reidemeister torsion, Ann. Math. **76** (1962), 137-147.

[MH]   J. Milnor and D. Husemoller, Symmetric Bilinear forms, Ergebnisse der mathematik und ihrer grenzegebiete 73, Springer-Verlag, 1973.

[Mo]   L. Moser, Elementary surgery on a torus knot, Pacific J. Math. **38** (1971), 737-745.

[N1]   P. E. Newstead, Topological properties of some spaces of stable bundles, Topology **6** (1967), 241-262.

[N2]   P. E. Newstead, Characteristic classes of stable bundles of rank 2 over an algebraic curve, Trans. A. M. S. **169**, 337-345 (1972).

[NZ]   I. Niven and H. Zuckerman, An introduction to the theory of numbers, 3rd ed., John Wiley and sons, 1972.

[Q]   D. Quillen, Determinants of Cauchy-Riemann operators over a Riemann surface, Funct. anal. appl. **19** (1985), 31-34.

[RG]   H. Rademacher and E. Grosswald, Dedekind sums, Carus mathematical monographs 16, M.A.A., 1972.

[Ro]   D. Rolfsen, Knots and links, Publish or perish, 1976.

[S]   J. Singer, Three dimensional manifolds and their Heegaard diagrams, Trans. AMS **35** (1933), 88-111.

[W1]   K. Walker, An extension of Casson's invariant to rational homology spheres, Bull. AMS **22** (1990) 261-268.

[W2]   K. Walker, The μ-invariant as a topological quantum field theory, in preparation.

[We]   A. Weinstein, Lectures on symplectic manifolds, C.B.M.S. regional conference series in math., no. 29, A.M.S., 1979.

BIBLIOGRAPHY

Milton Keynes UK
Ingram Content Group UK Ltd.
UKHW022110030124
435425UK00012B/670

9 780691 025322